Cambridge Eler

Elements in Current Archaeological To‹

edited by
Hans Barnard
University of California, Los Angeles
Willeke Wendrich
University of California, Los Angeles

ARCHAEOLOGICAL MAPPING AND PLANNING

Hans Barnard
University of California, Los Angeles

Shaftesbury Road, Cambridge CB2 8EA, United Kingdom

One Liberty Plaza, 20th Floor, New York, NY 10006, USA

477 Williamstown Road, Port Melbourne, VIC 3207, Australia

314–321, 3rd Floor, Plot 3, Splendor Forum, Jasola District Centre,
New Delhi – 110025, India

103 Penang Road, #05–06/07, Visioncrest Commercial, Singapore 238467

Cambridge University Press is part of Cambridge University Press & Assessment,
a department of the University of Cambridge.

We share the University's mission to contribute to society through the pursuit of
education, learning and research at the highest international levels of excellence.

www.cambridge.org
Information on this title: www.cambridge.org/9781009073240

DOI: 10.1017/9781009072069

© Hans Barnard 2023

This publication is in copyright. Subject to statutory exception and to the provisions
of relevant collective licensing agreements, no reproduction of any part may take
place without the written permission of Cambridge University Press & Assessment.

First published 2023

A catalogue record for this publication is available from the British Library.

ISBN 978-1-009-07324-0 Paperback
ISSN 2632-7031 (online)
ISSN 2632-7023 (print)

Additional resources for this publication at www.cambridge.org/barnard

Cambridge University Press & Assessment has no responsibility for the persistence
or accuracy of URLs for external or third-party internet websites referred to in this
publication and does not guarantee that any content on such websites is, or will
remain, accurate or appropriate.

Archaeological Mapping and Planning

Elements in Current Archaeological Tools and Techniques

DOI: 10.1017/9781009072069
First published online: May 2023

Hans Barnard
University of California, Los Angeles

Author for correspondence: Hans Barnard, nomads@ucla.edu

Abstract: This richly illustrated Element introduces the reader to the basic principles of archaeological mapping and planning. It presents both the mathematical and the practical backgrounds, as well as many tips and tricks. This will enable archaeologists to create acceptable maps and plans of archaeological remains, even with limited means or in adverse circumstances.

Keywords: archaeological survey, archaeological mapping, triangulation, leveling, total station

© Hans Barnard 2023

ISBNs: 9781009073240 (PB), 9781009072069 (OC)
ISSNs: 2632-7031 (online), 2632-7023 (print)

Contents

Figure 1 The eastern part of a map on papyrus showing gold mines, stone quarries, and other features in Wadi Hammamat, Egypt (*Museo Egizio*, Turin, Cat. 1879 + 1969 + 1899), drawn by Amennakhte, son of Ipuy, for Pharaoh Ramesses IV (1156–1150 BCE). This is generally considered the oldest surviving map. South is at the top, allowing the waters of the River Nile to flow down. Photograph courtesy of James Harrell.

He had bought a large map representing the sea,
Without the least vestige of land:
And the crew were much pleased when they found it to be
A map they could all understand.

The second stanza of *Fit the Second: The Bellman's Speech,* in: *The Hunting of the Snark: An Agony in Eight Fits* (1876).

Lewis Carroll (Charles L. Dodgson)

1 Theoretical Background: Points, Lines, Angles, and Polygons

Archaeology is special within the sciences because it is the only discipline that irretrievably destroys its evidence while it is being recovered. As the layers within an excavation unit are one by one removed and objects unearthed and retrieved, their all-important relative contexts can only be observed as this process progresses. It is impossible to reconstruct their intricate relationship and repeat the process. This places great responsibilities upon the excavators, who need to record the information that they infer as it becomes available. Ideally, their records should serve as a proxy for the excavation unit and allow future scholars to study it in ways similar to those used by the original excavators. Records comprise notes, photographs, drawings, and plans with abundant overlap, cross-references, and redundancies.These recording techniques, and more, do not render one another dispensable, rather, they are complementary.

Primary data on all archaeological features include their spatial and temporal properties. These are expressed in four dimensions: three in space – usually referred to as X or easting(s), Y or northing(s), and Z or elevation – and one in time, the age of the objects or their date of production or deposition. Temporal properties are usually established and recorded once the objects have been unearthed and moved into a laboratory to be cleaned, stabilized, analyzed, and studied. The spatial properties have to be recorded in situ, which potentially allows objects to remain where they were found, rather than be retrieved. This approach keeps the archaeological record intact and prevents issues related to storage, preservation, and ownership. This Element aims to introduce the basic principles of archaeological mapping and planning, which entails establishing, recording, and visualizing spatial data associated with archaeological features, ranging from ancient buildings to individual artifacts. It will not present detailed instructions on how to operate specific instruments or software packages, but rather the mathematical and practical backgrounds of mapping and planning in the field.

With this knowledge, archaeologists should be able to learn swiftly how to operate the instruments and software available to them, as well as assess the validity of the results that they obtain. This is especially relevant as access to state-of-the-art resources can be limited. Many of these are relatively expensive to purchase and maintain and they often have a rather short life span. This may not be an issue for well-established archaeological institutions, which are mostly based in Europe, but can form a significant obstacle for archaeologists elsewhere. Even when potentially available, many countries impose restrictions on the import or the use of electronic and imaging equipment, including electronic survey instruments and especially drones. Moreover, many archaeological sites are in places where access to the Internet or cell phone service, or even electricity, is intermittent or absent. Furthermore, most readily available technologies are designed to execute the reverse of archaeological mapping or planning, as they are primarily geared toward laying out designs in the field, rather than reducing reality into a map (Figure 1). Finally, during an initial visit to establish the archaeological potential of a larger area, it can be necessary to record observations of interest swiftly and with minimal means. The methods and techniques discussed in this Element, some of which date back centuries or even millennia, will enable archaeologists to create acceptable maps with simple means in adverse circumstances.

Like all practical skills, such as driving a car or playing the piano, mapping and planning cannot entirely be learned from an illustrated text, but rather by apprenticeship and practice. During this process, each archaeologist will develop a personal style and work flow, which nevertheless should always aim to reach a final result that can readily be understood and used by other archaeologists. The most important skill that an archaeological surveyor should develop, or have, is the ability to visualize a mental image of the area to be planned and imagine what the final map should look like. Making a sketch map before starting the actual survey work will not prove just helpful in this, but near indispensable. Additional useful skills include a constant awareness of the cardinal directions and the length of one's stride. All these will improve with experience and enable a proficient surveyor to create a relatively accurate sketch map in a short amount of time. All digital methods and technologies are complementary to these basic skills and cannot fully replace them as an effective avenue to gain insights in exposed ancient remains. The importance of knowing the basics and the value of drawings can hardly be overstated. They allow for interpretations to be reflected – by highlighting or instead disregarding specific details – and, more importantly, they provide an opportunity to study the ancient remains and their intricate relationships to the extent necessary to produce an accurate drawing.

Examples of different types of maps and plans are provided throughout the text – starting with Figure 2, but mostly concentrated in Section 6 – to serve as examples and provide inspiration. In other places specific values are listed in the text or tables. These are provided to complement the illustrations and the text, as they can readily be obtained in much more detail from a calculator or cell phone. Where necessary these numbers and the equations associated with them will be embedded in electronic survey equipment and software. A list of the most often used abbreviations is provided in Table 16.

All measured survey activities are based upon points, lines, and polygons, almost exclusively triangles. To provide a handle on these and allow their properties to be expressed, established, or calculated, these are located within

Figure 2 Photograph of the remains of ancient structures, looking south (top), and the corresponding measured plan of the same structures (bottom)

a grid or coordinate system. A line is the shortest connection between two points and can be imagined to continue infinitely in either or both directions (Figure 3, top). Lines are one dimensional, meaning that they have no width or height, but

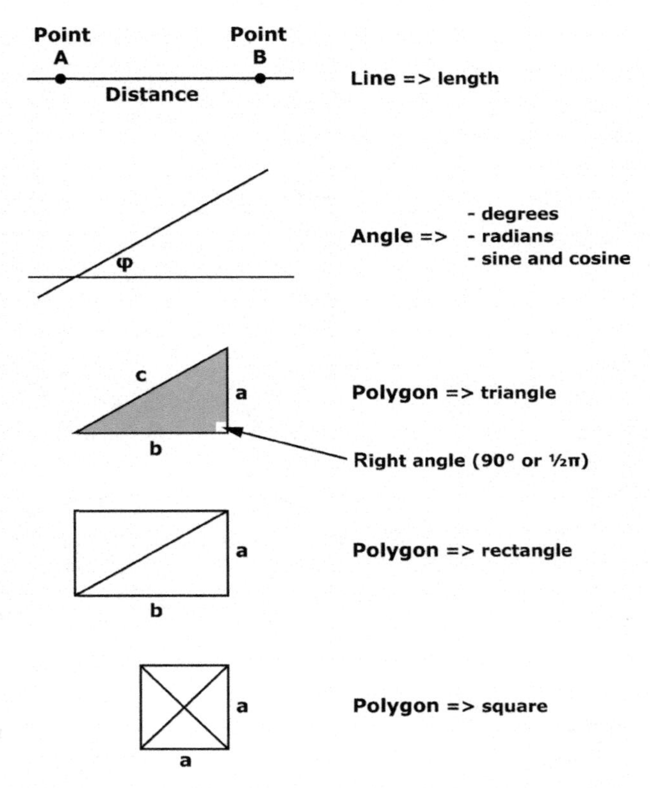

Figure 3 The basic elements of mapping and planning: points, lines, angles and polygons (triangles)

only length. A distance between two points is the length of the line between these points. In archaeology distances are expressed in millimeters, centimeters, meters, and kilometers (Table 1), as they are in other sciences. One meter is 39.37008 inches or 3.28084 feet (Table 2). Note how metric units can be converted into each other by simply moving the decimal point three places, either to the left or the right.

When drawing a measured plan or map of an archaeological site, or any other feature in the terrain, this involves uniformly scaling down the measured distances to fit the selected medium and purpose of the final result. For this distances are divided by a convenient number, most often 10, 20, or 100 (written as 1:10, 1:20, and 1:100, respectively), or multiples thereof (Table 3). For practical purposes it is better to avoid scales that will result in awkward numbers, such as 3 and 30, but also 4 and 25. The topographic maps that exist of most regions in the world and are often the base for more detailed archaeological maps are most commonly drawn to a scale of 1:25,000 (1 cm on the map is 250 m in the terrain and 1 km in the terrain is 4 cm on the map) or 1:50,000

Table 1 Terminology of the metric system

Decimals	10^n	Prefix	Symbol
1 000 000 000 000 000 000	10^{18}	exa-	E
1 000 000 000 000 000	10^{15}	peta-	P
1 000 000 000 000	10^{12}	tera-	T
1 000 000 000	10^{9}	giga-	G
1 000 000	10^{6}	mega-	M
1 000	10^{3}	kilo-	k
1	10°		
0.001	10^{-3}	milli-	m
0.000 001	10^{-6}	micro-	µ
0.000 000 001	10^{-9}	nano-	n
0.000 000 000 001	10^{-12}	pico-	p
0.000 000 000 000 001	10^{-15}	femto-	f
0.000 000 000 000 000 001	10^{-18}	atto-	a

Table 2 Conversions of selected units of distance

Unit	Equivalent		
1 millimeter	0.03937 inch		
1 meter	39.37008 inches	3.28084 feet	
1 kilometer	0.62137 mile	3280.83990 feet	39 370.07874 inches
1 inch	25.4 millimeter		
1 foot	304.8 millimeters	0.3048 meter	
1 yard	914.4 millimeters	0.9144 meter	
1 mile	1 609 344 millimeters	1609.344 meters	1.60934 kilometers

Table 3 Length in centimeter on a map of selected distances in the terrain at the most commonly used scales. Note that specific scales are more appropriate for certain ranges of features than others

	Scale				
Actual distance	1:10	1:20	1:100	1:200	1:1,000
1 cm	0.1	–	–	–	–
10 cm	1	0.5	0.1	–	–
1 m	10	5	1	0.5	0.1
10 m	100	50	10	5	1
100 m	–	–	100	50	10
1 km	–	–	–	–	100

(1 cm on the map is 500 m in the terrain and 1 km in the terrain is 2 cm on the map). Maps showing larger areas usually have too little detail to be of much use for archaeological research.

Lines that are not exactly parallel will eventually intersect. This can create either four angles that are exactly the same, identified as right angles, or a pair of obtuse and a pair of acute angles (Figure 3). Angles can be expressed in three different, but related ways. All involve imagining the meeting point of the two lines to be the center of a circle with a radius of one measuring unit (which can be 1 millimeter, inch, foot, yard, meter, kilometer, mile, or something else entirely), a so-called unit circle. The actual size of the unit circle is irrelevant as during the scaling of distances angles will remain unchanged.

The first method to express the size of an angle is to match it with the segment of the circle created by the two lines (Figure 4). Conventionally, a circle is divided into 360° (degrees), each of which is subdivided into 60′ (minutes), each of which is again divided into 60″ (seconds), or $60 \times 60 = 3600''$ for each degree (Animation 1). Alternatively, degrees can be divided into minutes and decimal minutes (D° M.mmmmm′, Table 4) or into decimal degrees (D.ddddd°). In this system, a right angle is 90° (or 270°) and a straight line 180°. Other methods to divide circles have been developed, but are much less common. Most often used is the division of the full circle into 400 gons (400^g), making a right angle 100^g (or 300^g) and a straight line 200^g.

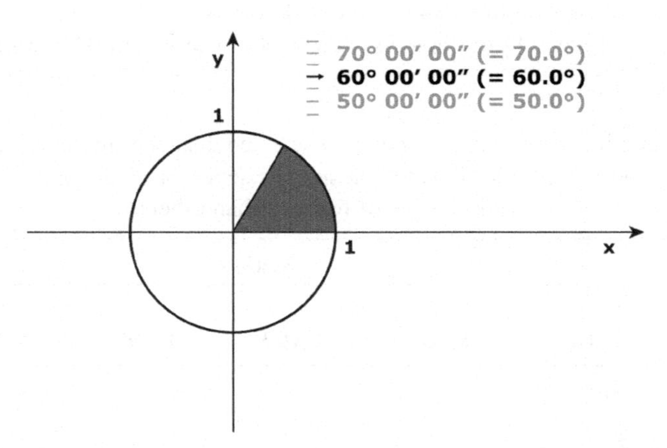

Animation 1 The unit circle (with a radius of one unit of length), divided into 360°. The animated version of the image is available at www.cambridge.org/barnard

Table 4 Conversion of selected minutes and decimal minutes. These figures are provided here to complement the text and illustrations

Minutes	Decimal minutes	Decimal minutes	Minutes
0′	0	0	0′
10′	D.16666 . . .	D.10	6′
15′	D.25	D.20	12′
20′	D.33333 . . .	D.30	18′
30′	D.5	D.40	24′
40′	D.66666 . . .	D.50	30′
45′	D.75	D.60	36′
50′	D.83333 . . .	D.70	42′
60′ (1°)	D+1	D.80	48′
		D.90	54′
		D+1	60′ (1°)

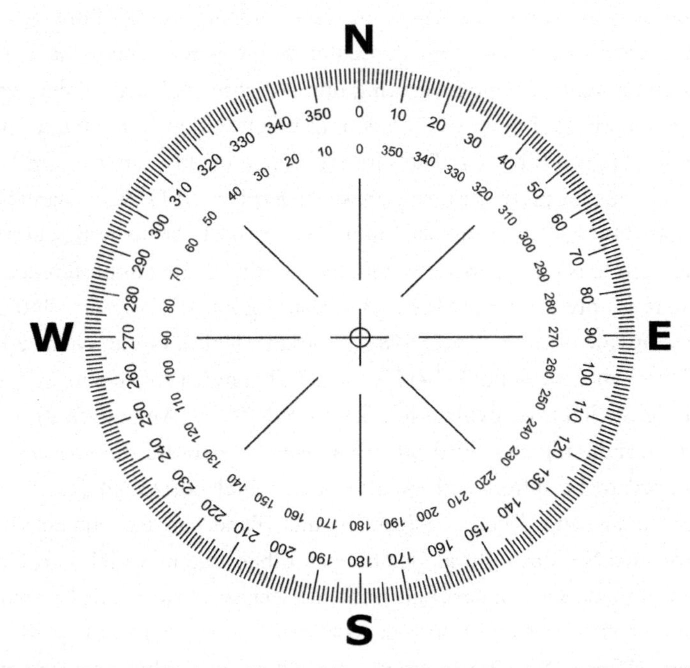

Figure 4 A full circle divided into 360° (degrees), see Figure 28

Table 5 Conversion of selected angles into radians, see Figure 8. These figures are provided here to complement the text and illustrations

Degrees	Radians	
0°	0	0
10°		0.17453
20°		0.34907
30°	⅙ × π	0.52360
40°		0.69813
45°	¼ × π	0.78540
50°		0.87266
60°	⅓ × π	1.04720
70°		1.22173
80°		1.39626
90°	½ × π	1.57080

Another way to express the size of an angle is by the length of the segment of the unit circle – the circle with its center at the place where the two lines intersect and a radius of one measuring unit – that is enclosed by the two lines that make that angle (Figure 8). This length is referred to as the radian (rad) of that specific angle (Table 5). The circumference of any circle is $2\pi \times R$, or 6.28318 . . . × the radius of the circle, in which π (pi) = 3.14159 . . ., which can be approximated by 22/7 or, more accurately, by 355/113. As the radius in the case on the unit circle is one measuring unit, by definition, the circumference of the full circle is simply 2π (= 6.28318 . . .) measuring units. Therefore, 360°, a full circle, is equivalent to 2π or 6.28318 . . . radians. Half a circle, enclosed by an angle of 180°, measures π or 3.14159. . . . A right angle (90°) encloses a quarter of the circle, which is equivalent to π/2 or 1.57079 . . . (Animation 2).

Before introducing the third and final way to express angles, coordinate systems need to be introduced. A grid is a virtual horizontal plane laid out across the landscape. Every position in this plane can be expressed by its distance to two lines at right angles to each other, one of which is referred to as the X-axis (*abscissa*), or easting(s), and the second as the Y-axis (*ordinate*), or northing(s). The X-axis and Y-axis cross at right angles in a single point, called the Origin (Figure 5). The position of each point within the grid can be expressed by (x, y) coordinates, so-called Cartesian coordinates (Figure 6,

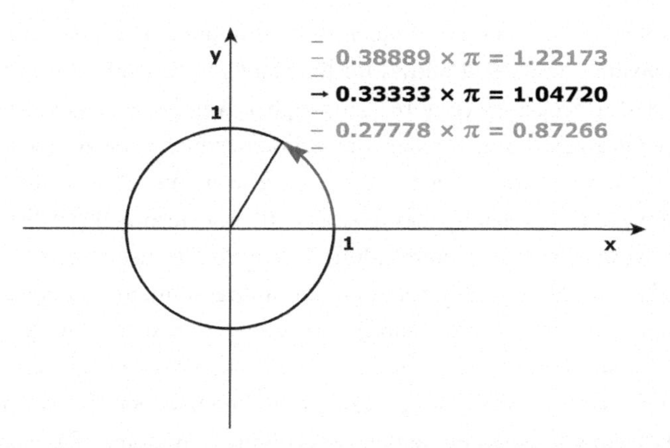

Animation 2 The unit circle (with a radius of one unit of length), divided into 2π (= 6.28318 . . .) radians. The animated version of the image is available at www.cambridge.org/barnard

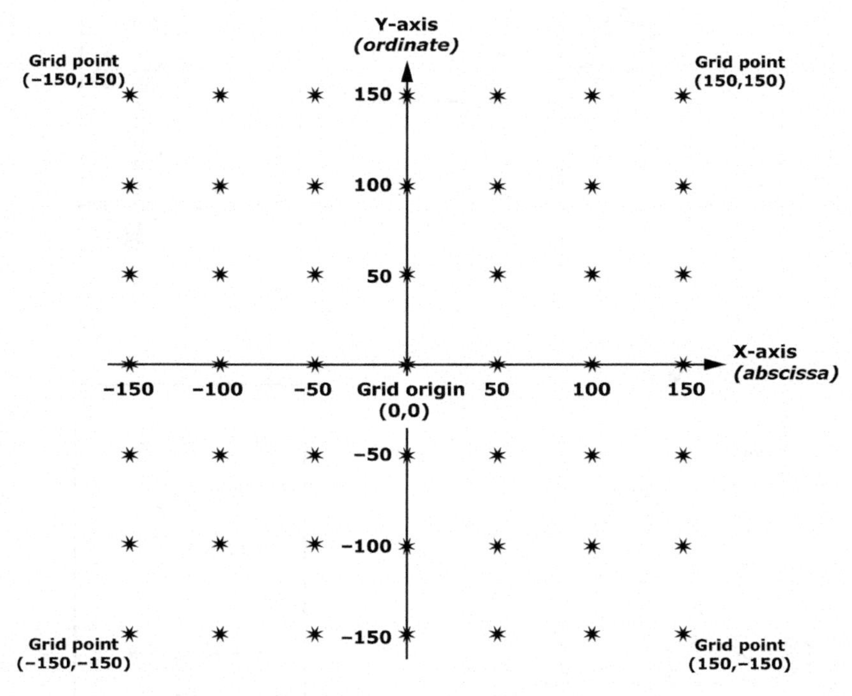

Figure 5 The Cartesian coordinate system, see Figure 13

top-right), named after the French philosopher and scholar René Descartes (1596–1650). Cartesian coordinates can be measured directly, especially at small scales in a relatively flat and open terrain, with one or more tape measures.

In archaeology, this method is frequently used to document objects and layers within excavation units as it allows for the results to be evaluated immediately and corrected if necessary (Figure 7, top-right). Another often used method to locate a find in an excavation unit is triangulation, which involves measuring the distance of each find to two or three known grid points, almost invariably the corners of the excavation unit, and subsequently transferring these distances to a scaled drawing (Figure 7, bottom-left). To locate is used here to mean to establish the coordinates of a specific point in three-dimensional space.

A second way to express coordinates within a grid is closely related to triangulation and comprises measuring the distance between an unknown and a known grid point, as well as the angle of the line between the two points and the X-axis, or any line parallel to that axis (Figure 6, bottom-left). Apart from a

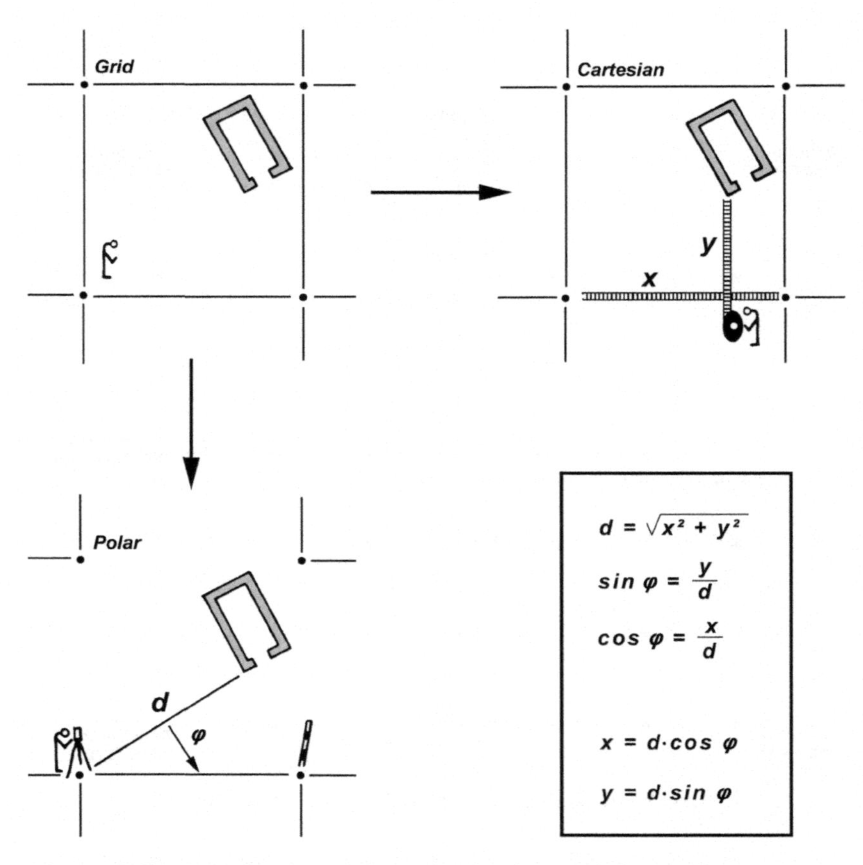

Figure 6 The Cartesian (top-right) and polar (bottom-left) coordinate systems and the formulae to convert one into the other (bottom-right), see Figures 2 and 9

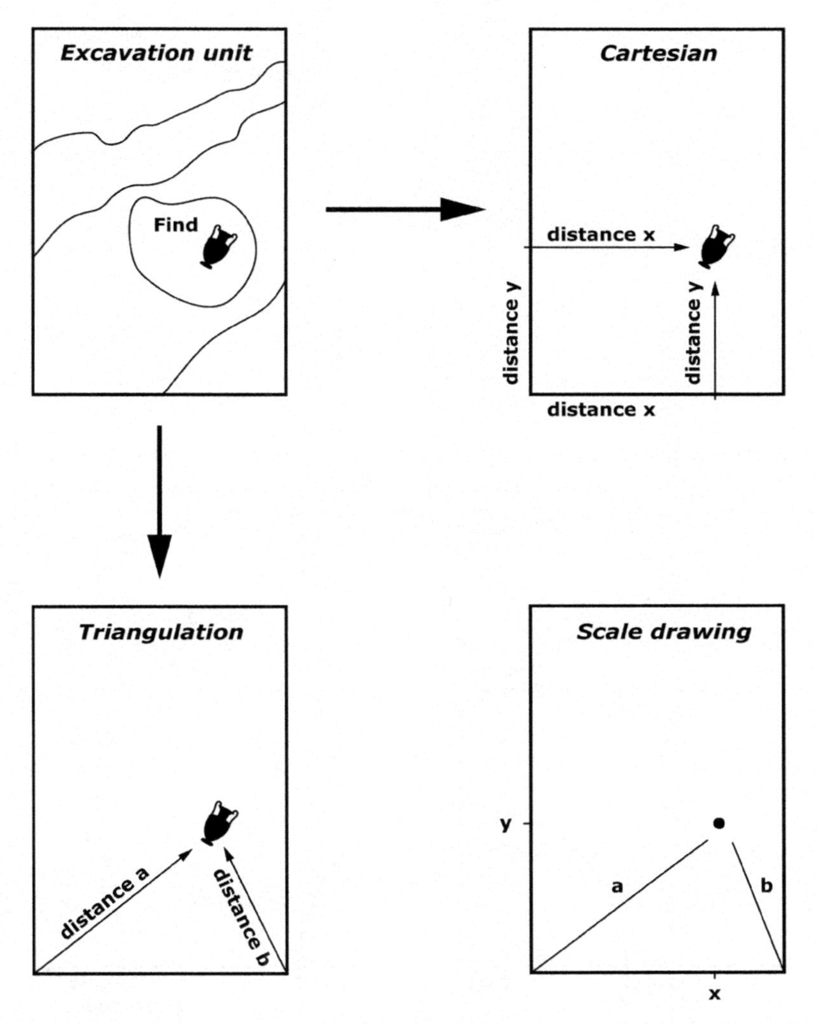

Figure 7 Two methods of locating finds within an excavation unit: Cartesian (top-right) and triangulation (bottom-left)

distance measuring device, this method requires an instrument capable of accurately measuring angles. When drawing polar coordinates, distances are scaled down while angles remain unchanged. Apart from being used to create a scaled drawing, however, the measured angles and distances can also be entered into formulae that allow the calculation of a variety of other angles and distances. For this, a third method to express angles, other than degrees and radians, is deployed.

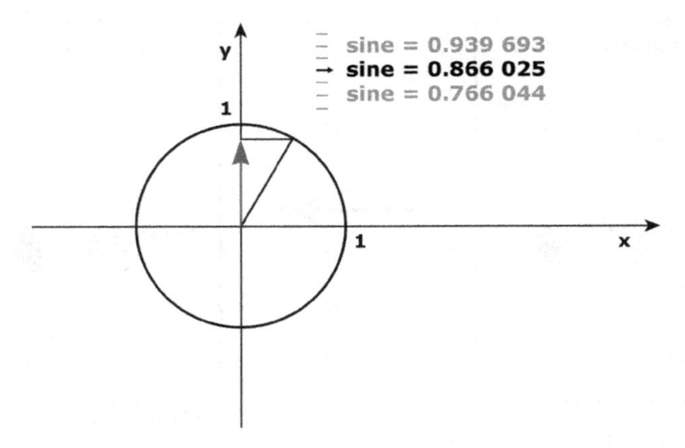

Animation 3 The relationship between angles and their sine (y-coordinate). The cosine is the corresponding x-coordinate. The animated version of the image is available at www.cambridge.org/barnard

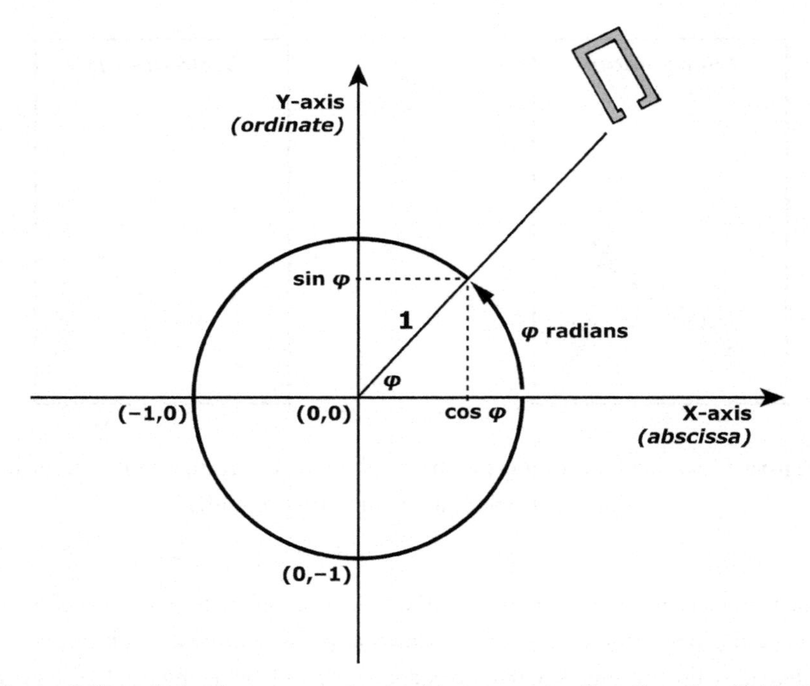

Figure 8 The relation between angle φ and its expression in radians (see Table 5), and its sine and cosine (see Table 6)

A third method to express angles is by imagining one of the lines that make the angle as the X-axis of a local coordinate system and express the angle as the x- or y-coordinate of the intersection of the unit circle and the second line that creates the angle (Figure 8). The x-coordinate is referred to as the cosine of the enclosed angle and the y-coordinate as its sine (Animation 3). The sine and cosine of any angle obviously vary between –1 and 1, while $(\sin \varphi)^2 + (\cos \varphi)^2 = 1$ as $a^2 + b^2 = c^2$ (theorem of Pythagoras). This method allows the easy transformation of polar into Cartesian coordinates, and vice versa (Figure 5, bottom-right). The combination of coordinates, where the position of each point can be expresses by its (x, y) coordinates, and the sine and cosine properties of angles allows for the relative position of any point within the grid to be calculated (Figure 9). The sines and cosines of angles can be found with the help of a calculator or a table (Table 6). As obvious from Figure 6 and Figure 9, the values of sines and cosines are directly related to the lengths of the sides of triangles with one right angle ($90° = \pi/2$ or 1.57079 radians).

The final basic element of all mapping and planning activities – after points, coordinates, lines, and angles – is the polygon. A polygon is a figure that is enclosed by a finite number of straight, end-to-end connected line segments. Examples of polygons are triangles, rectangles (squares), pentagons, hexagons, heptagons, and octagons, with three, four, five, six, seven, and eight line segments, respectively, meeting in as many angles. All polygons can be reduced to a combination of a number of triangles, actually two triangles less than the total number of the sides (or angles) of

Table 6 Sines and cosines of selected angles, see Figure 8. These figures are provided here to complement the text and illustrations

	Angle		
Sine	**Degrees**	**Radians**	**Cosine**
0	0°	0	1
0.17364 . . .	10°	0.17453	0.98480 . . .
0.34202 . . .	20°	0.34907	0.93969 . . .
0.5	30°	0.52360	0.86602 . . .
0.64278 . . .	40°	0.69813	0.76604 . . .
0.70710 . . .	45°	0.78540	0.70710 . . .
0.76604 . . .	50°	0.87266	0.64278 . . .
0.86602 . . .	60°	1.04720	0.5
0.93969 . . .	70°	1.22173	0.34202 . . .
0.98480 . . .	80°	1.39626	0.17364 . . .
1	90°	1.57080	0

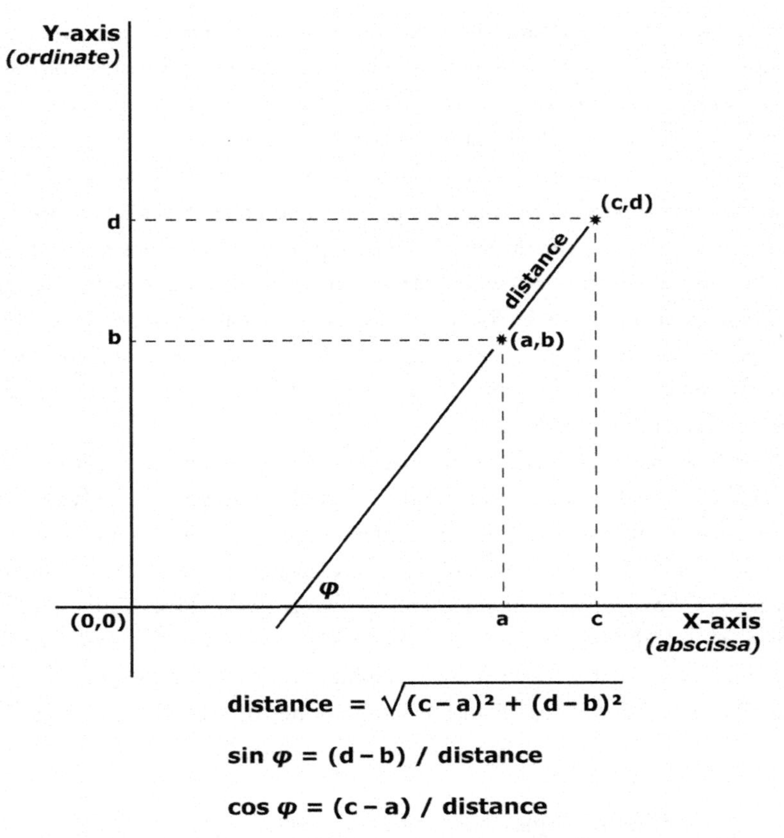

Figure 9 Basic calculations enabled by Cartesian coordinates, see Figure 6. Electronic survey equipment and software will have these equations embedded

the polygon. This fact places triangles and all their specific properties central in all survey work. There are a number of special triangles, including the equilateral, the isosceles, and the right-angled triangle. Equilateral triangles have three sides of the same length and three angles of 60° (= π/3 or 1.04719 radians), while isosceles triangles have two sides of equal length and two angles that are the same. The sum of the three angles of all triangles equals 180° (= π or 3.14159 radians).

A right-angled triangle has one angle of 90° (= π/2 or 1.57079 radians). The side opposite the right angle, the longest of the three sides, is referred to as the hypotenuse. The relationship between the length of the hypotenuse and the lengths of the two sides that meet at the right angle is defined by the theorem of Pythagoras: $a^2 + b^2 = c^2$ (in which a and b represent the lengths of the two sides creating the right angle and c is the length of the hypotenuse). The theorem of Pythagoras can obviously be used to calculate the lengths of the sides of a right-angled triangle, but also to create a right angle by laying out the lengths of the three sides that satisfy the theorem. For instance, two lines with a length of 4 and 6 units, respectively, will meet at a right angle if their other ends are connected by a hypotenuse of $\sqrt{(4^2 + 6^2)}$ units, which equals $\sqrt{(16 + 36)} = \sqrt{52} = 7.21110$ units. In other words, a rectangle with sides of 4 and 6 units long, respectively, will have two diagonals of 7.21110 units (Figure 11).

There are two ways to simplify this method of creating right angles and thus rectangles and squares (Figure 10). The first is to create a square in which b = a, transforming the theorem of Pythagoras into $a^2 + a^2 = c^2$, or $2 \times a^2 = c^2$. Taking

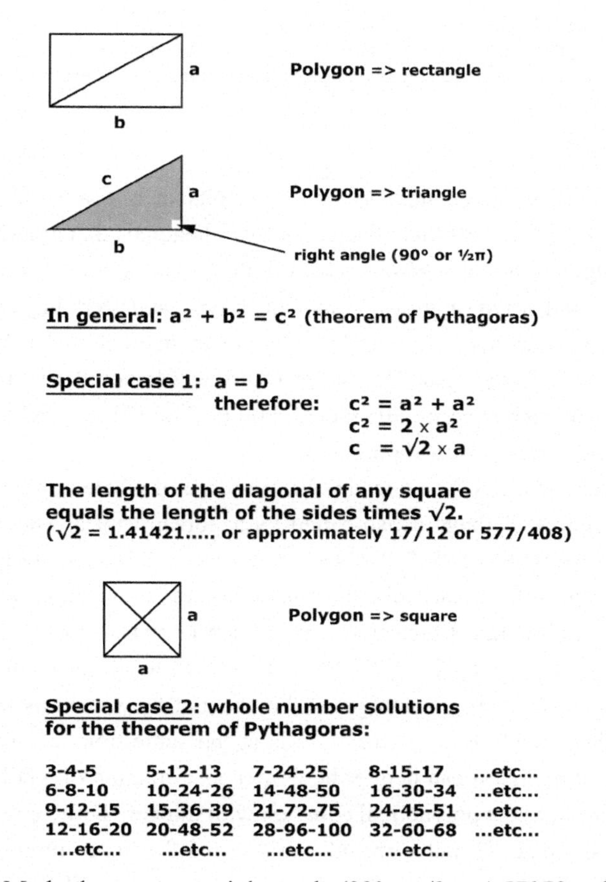

Figure 10 Methods to create a right angle ($90° = \pi/2$ or 1.57079 radians) using the theorem of Pythagoras: hypotenuse = $\sqrt{(a^2 + b^2)}$

Table 7 Length of the diagonals of selected squares

Sides	Diagonal	
1 × 1	1 × √2	1.414 . . .
2 × 2	2 × √2	2.828 . . .
3 × 3	3 × √2	4.242 . . .
4 × 4	4 × √2	5.656 . . .
5 × 5	5 × √2	7.071 . . .
6 × 6	6 × √2	8.485 . . .
7 × 7	7 × √2	9.899 . . .
8 × 8	8 × √2	11.313 . . .
9 × 9	9 × √2	12.727 . . .
10 × 10	10 × √2	14.142 . . .
15 × 15	15 × √2	21.213 . . .
20 × 20	20 × √2	28.284 . . .

the square root of either side, to find c, results in $c = a \times \sqrt{2}$, in which $\sqrt{2} = 1.41421$. This means that all squares have diagonals that are $\sqrt{2} = 1.4142$ times the length of the sides of the square (Table 7). This method was apparently used to lay out right angles in ancient Egypt and certainly in ancient Mesopotamia as evident from Babylonian mathematical tablet YBC 7289 kept in the Yale Babylonian Collection (Sterling Memorial Library). In the latter, √2 is approximated by 305470/216000 (= 1.41421 . . .), which is more than sufficient for all practical purposes.

The second avenue is facilitated by the existence of whole number solutions for the theorem of Pythagoras, so-called Pythagorean triples. The triple most often used in the field is 3–4–5 ($3^2 + 4^2 = 5^2$ as $9 + 16 = 25$), or a multiple thereof (Figure 10, bottom). Remarkably, this method seems to have not been used in ancient times, although it appears to have been known to the ancient Greek scholar Proclus Lycaeus (412–485 CE), and probably much earlier. Both √2 and π are irrational numbers, meaning that they cannot be written as fractions of whole numbers, such as ¼, ½, or ¾; π is at the same time a transcendental number, meaning that it can neither be written as a fraction of whole numbers nor as the solution of a polynomial equation with integer coefficients. The only way to approximate √2 and π are infinite decimal fractions, although various whole number fractions have been used in the past and are still sometimes used for convenience (Figure 11).

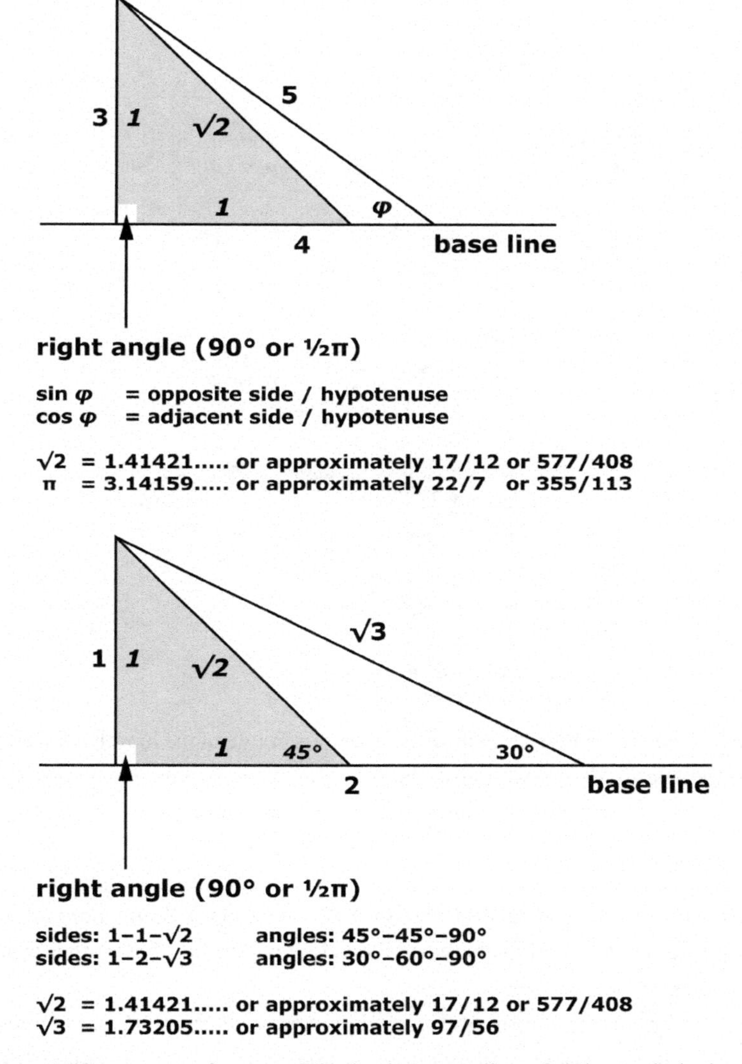

Figure 11 Summary of some of the basic properties of right-angled triangles that are of importance to archaeological fieldwork

A right-angled triangle with sides of 1–1–√2 is obviously at the same time an isosceles triangle, in which two of the angles are necessarily the same. Given that the sum of the three angles has to be 180° (= π or 3.14159 radians), these two angles are 45° (= π/4 or 0.78539 radians) each (Figure 11, top). A final right-angled triangle of note is the triangle with sides of 1–2–√3 (√3 = 1.73205, which can quite accurately be approximated by 97/56). This triangle has one right angle (Figure 11, bottom), one angle measuring 60° (= π/3 or 1.04719 radians),

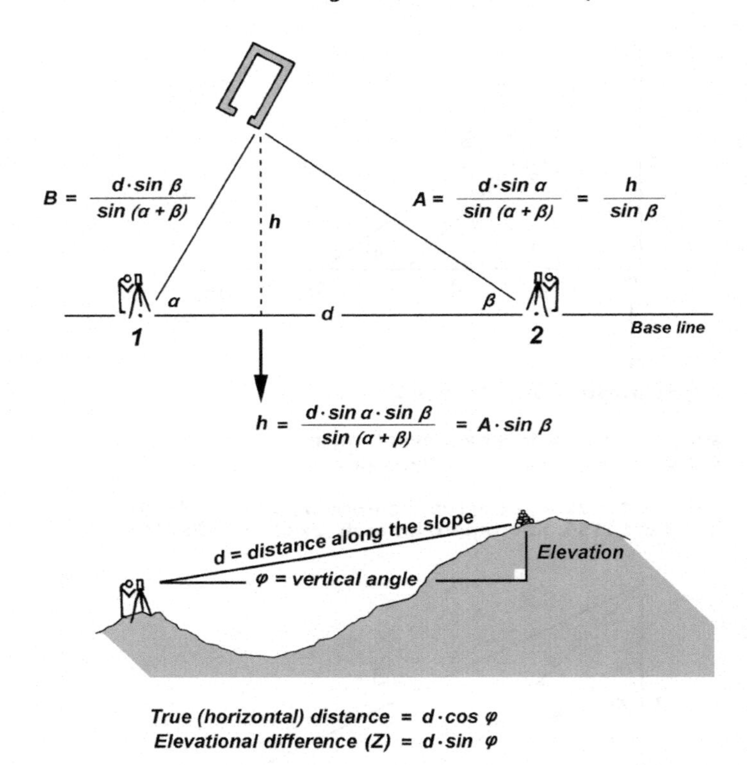

Figure 12 The formulae that define the relations between the lengths of the sides and the enclosed angles of any triangle. Electronic survey equipment and software will have these equations embedded

and a third angle measuring 30° (= π/6 or 0.52359 radians). Some more involved formulae define the relations between the sides and angles of any other triangle (Figure 12).

The surface area of polygons can be expressed in m^2 (square meters; 1 m^2 = 1 m × 1 m) or km^2 (square kilometers; 1 km^2 = 1 km × 1 km), although in archaeology hectares (ha) are often used. One hectare is the area of a 100 m × 100 m, or 10,000 m^2 (Table 8).

2 Theoretical Background: The Third Dimension

A Cartesian grid is defined as a virtual horizontal plane in which every position can be expressed by its distance to the X- and the Y-axes: (x, y) or E–N. The properties of angles and triangles allow for the relations between two points with known coordinates to be calculated (Section 1). In archaeological mapping and planning, however, two factors complicate this principle, as they do in all survey work. The first is that

Table 8 Surface area of selected squares and rectangles

	10 m	**100 m**	**1,000 m**
10 m	100 m²	1,000 m²	10,000 m² 1 ha 2.47105 acres
100 m	1,000 m²	10,000 m² 1 ha 2.47105 acres	100,000 m² 10 ha 24.71054 acres
1,000 m	10,000 m² 1 ha 2.47105 acres	100,000 m² 10 ha 24.71054 acres	1000,000 m² 1 km² = 100 ha 0.38610 square miles

the topography of the landscape will result in most points of interest being either above or below the virtual horizontal plane of the grid. The second is that it is impossible to reduce the three-dimensional, almost spherical curvature of the earth into a two-dimensional horizontal surface. For the relatively short distances across which archaeological mapping and planning usually takes place, rarely more than 10 km or 5 miles, and most often much less, the latter can usually be ignored. The former, however, is usually relevant, especially when excavating. The difference between the actual position of any given point and the Cartesian plane is referred to as its elevation and is usually expressed as its z-coordinate (Figure 13). The z-coordinate can either be measured directly or calculated (Figure 12, bottom).

In order to measure the elevation of a point directly, an artificial horizon needs to be created at a known elevation. This elevation can be known from previous survey work, established by a barometric altimeter, or by a device that can calculate its position with the help of the signal from a constellation of satellites (Section 4). It can also be an arbitrary number (for instance 100 or 1,000 m), allowing for a single site to be mapped with internal consistency. The most direct way to create an artificial horizon across a limited distance (less than 10 m), such as an excavation unit, is to pull a sting from a point with a known elevation (usually a temporary bench mark) and keep this in a horizontal position with the aid of a line level, which is a simple spirit level that can be suspended from a string (Figure 14). As the elevation of the string is known and the same in any direction, provided the string is kept horizontal as indicated by the line level, the elevation of any point below the string can simply be measured with a tape measure or a similar measuring device (Figure 15).

The second and most accurate way to establish the elevation (altitude) of points across the landscape requires an instrument called a level, or dumpy level

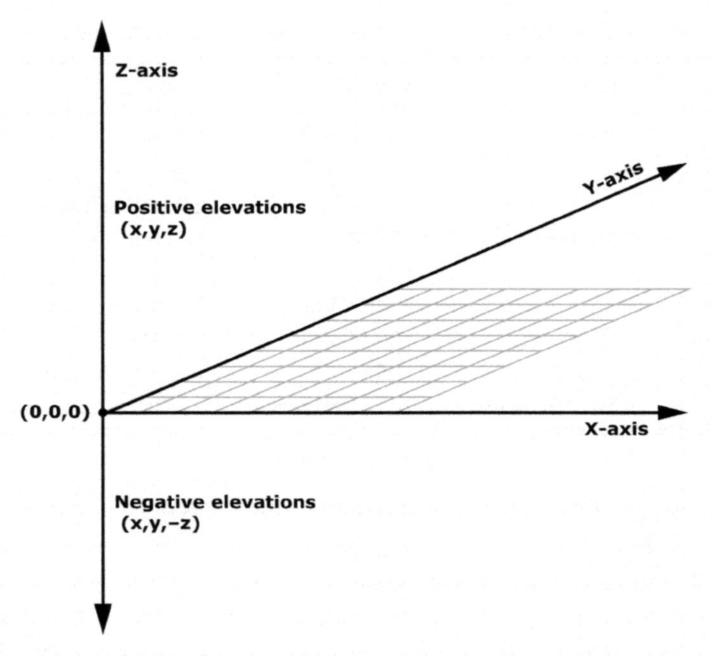

Figure 13 The three-dimensional Cartesian coordinate system (x, y, z)

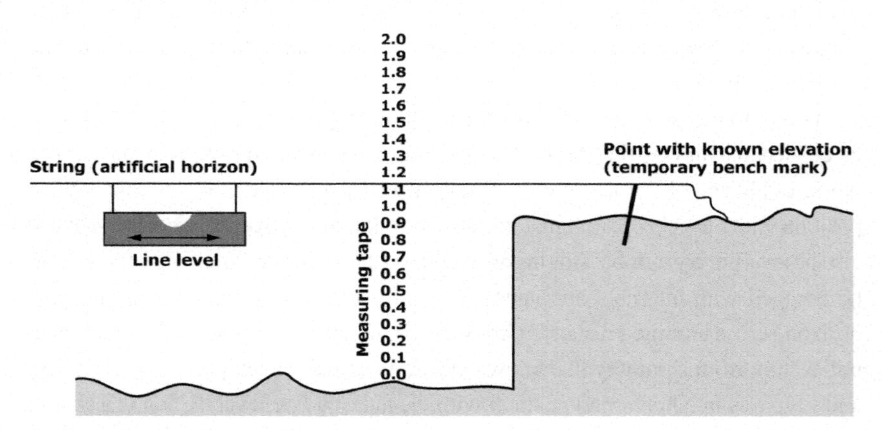

Figure 14 Measuring elevations over short distances can be done directly by using a line level suspended from a string, to create an artificial horizon, and a tape measure, to measure the distance between the unknown point(s) and the known horizontal plane, see Figure 15. Here, the point of interest is 1.14 m below the temporary bench mark

Figure 15 Archaeologists establish the elevation of an archaeological deposit with the aid of a line level and a measuring tape, see Figure 14

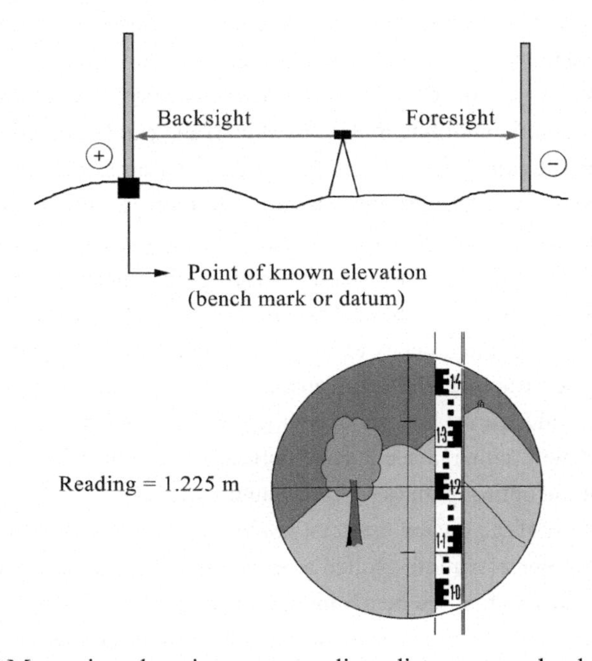

Figure 16 Measuring elevations over medium distances can be done directly with a level instrument (dumpy level), to create an artificial horizon (top), and a stadia rod (bottom), to measure the distance between the unknown point(s) and the known horizontal plane, see Figures 17 and 18. Here, the artificial horizon is 1.225 m above the bench mark (BM + 1.225 m)

Video 1 Field archaeologists demonstrate how to establish the Z-coordinate (elevation) of selected points. Video available at www.cambridge.org/barnard

(Figure 16), used in combination with a calibrated pole, usually 4 m tall, referred to as a stadia rod (Figure 17). A level is a telescope with an attached spirit level and mounted on a plate with three or four screws with which the plate and telescope can be maneuvered into a perfectly horizontal position. As the telescope has a horizontal line in its optical path and can be turned a full circle along its vertical axis, a level instrument creates an optical artificial horizon, which can be viewed through the telescope. Before use, the eyepiece of the telescope should be adjusted until this line pops into crisp focus. A different control on the instrument will focus the image of the landscape. It is better to work with the instrument without (sun)glasses and rely on the optics of the telescope. The instrument needs to be set up upon a steady tripod placed in a strategic position from which the full area to be surveyed can be seen (Video 1). To help secure an accurate horizontal position of the telescope most instruments are equipped with either a secondary, optically split spirit level, or a prism suspended in the optical path (a so-called automatic level).

The elevation of the optical artificial horizon can be calculated from a fixed point with a known elevation – called a bench mark, datum, or monument – by placing the stadia rod on the bench mark and adding the reading on the stadia rod to the elevation of the bench mark. Such looking back to a known point is referred to as taking a backsight. If, for instance, the reading on the stadia rod is 1.225 (Figure 16), the elevation of the horizon is 1.225 m higher than the elevation of the bench mark. This is often referred to as the day height or simply the height of the instrument (HI = BM + 1.225 m). The elevation of unknown

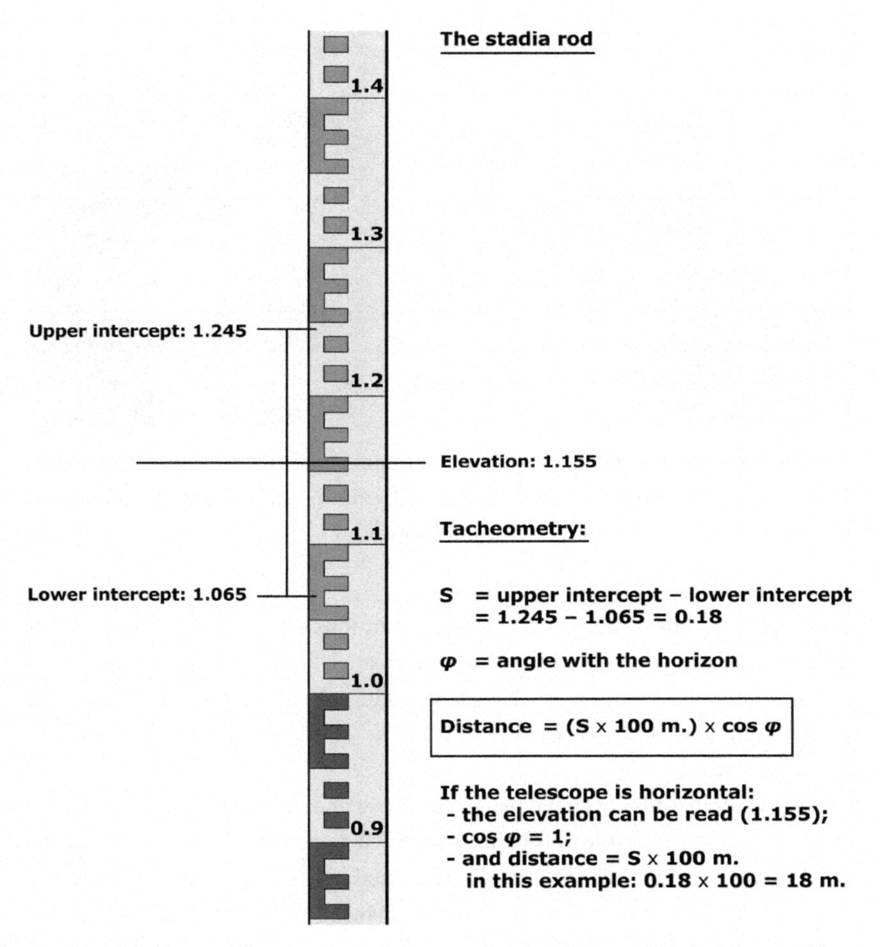

Figure 17 The usual layout of a stadia rod, divided into cm blocks, grouped into positive and negative E-shaped sets of five. The crosshairs in the telescope of the level instrument (dumpy level) indicate where the optical artificial horizon intersects the stadia rod, see Figures 16 and 18. Here, the point of interest is 1.155 m below the artificial horizon (HI – 1.155 m). The distance between the instrument and the rod can be estimated by tacheometry

points can now be established by taking as many foresights as there are points to be measured (Figure 18). For each, the stadia rod is placed on the point of interest and read, after which the reading is subtracted from the height of the instrument. As the optical artificial horizon created by the level instrument has to intersect the stadia rod, the instrument always has to be higher than the bench mark (and the reading on the stadia rod thus always added) as well as the points

Figure 18 Archaeologists establish the elevation of an archaeological deposit with the aid of a level instrument (dumpy level) and a stadia rod, see Figures 16 and 17

of interest (and the readings on the stadia rod thus always subtracted). If, for instance, the elevation of the artificial horizon is 1492.5 m above sea level (mASL), and the reading on the stadia rod is 1.155 (Figure 16), then the elevation of that point is 1492.5 − 1.155 = 1491.345 mASL (elevation = HI – 1.155 m).

Many level instruments have markings above and below the line marking the artificial horizon that allow for the distance between the instrument and the stadia rod to be estimated (Figure 17), a method referred to as tacheometry and in a modified form already used by the Romans. To estimate the distance the difference between the two readings is simply multiplied by 100 m. If the level instrument allows for a horizontal angle to be set and measured, this enables to creation of simple maps (Section 1), or the introduction of contour lines on existing maps (Figure 20). It is also useful for any surveyor to know the length of her or his stride, allowing distances to be estimated fairly confidently by counting paces. Many of the archaeological maps created by John Gardner Wilkinson (1797–1875), for instance, were measured in this fashion. When an instrument is used with a telescope that can also be rotated along the horizontal axis to measure vertical angles, such as a transit or a theodolite, the difference between the two readings needs to be multiplied by 100 m times the cosine of the angle with the horizon (Figure 17).

Establishing the accurate elevation (altitude) of one or more new bench marks when starting survey work in a new area can be challenging. It may be possible to use the results of previous work done nearby and measure or calculate the elevation of the new bench marks from existing points with a known elevation. Depending on the distance and the condition of the terrain between the known and unknown points, this may require some time to achieve. If the area of interest is close to a sea or ocean, the water level can be used as a relatively accurate approximation of the average ocean level. In other cases, it might be sufficient to assign an arbitrary elevation (for instance 100 or 1,000 m) to a central bench mark, allowing for a site to be surveyed with internal consistency. The resulting map can then later be combined with a topographic map of the area and all elevations amended to fall closer to their actual value.

Two methods are available to independently measure the elevation of points far from bench marks with a known elevation. The first is with the help of a barometric altimeter, which uses the fact that the air pressure, the weight of the column of air above the observer, drops almost linearly when moving to higher altitudes (Figure 19, Table 9). The actual value of the air pressure, however, also depends on the local temperature and weather. The altimeter thus has to be

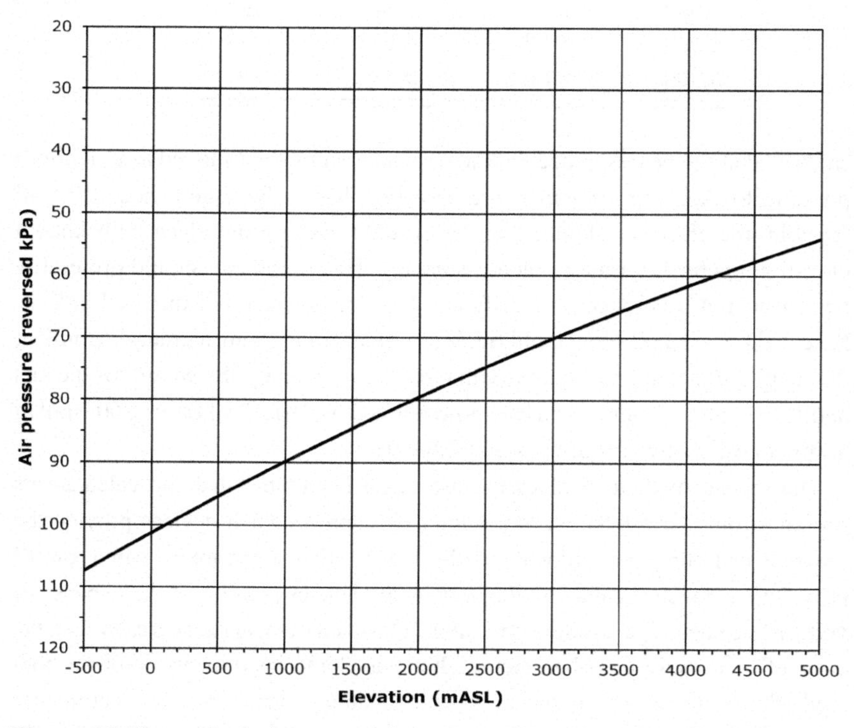

Figure 19 Theoretical air pressure between –500 and 5000 mASL, assuming an air pressure at sea level of 101.325 kPa (1 atmosphere) and a temperature of 15°C (60°F). Note the reversed vertical axis, see Table 9

Table 9 Theoretical air pressure at selected elevations above sea level in kilo-Pascal (kPa), atmosphere (atm) and mm mercury (mmHg = torr), assuming an air pressure at sea level of 1 atmosphere = 101.325 kPa and a temperature of 15°C (60°F), see Figure 19. These figures are provided here to complement the text and illustrations

Elevation	kPa	atm	mmHg
−500	107.5	1.06	806
0	101.3	1.00	760
500	95.5	0.94	716
1,000	89.9	0.89	674
1,500	84.6	0.83	634
2,000	79.5	0.78	596
2,500	74.7	0.74	560
3,000	70.1	0.69	526
3,500	65.8	0.65	493
4,000	61.6	0.61	462
4,500	57.7	0.57	433
5,000	54.0	0.53	405

calibrated at the nearest place with a known elevation and moved as quickly as possible to the point of which the elevation has to be established. If at all feasible, the altimeter should then be brought back to the place with known elevation to check for major inconsistencies. This round trip should preferably be repeated at least once on a different day. Air pressure is expressed in kilo-Pascal (kPa = 1,000 N/m^2), although the older units "mm mercury" (mmHg, the height of a column of mercury that can be held up by the weight of the air) and "atmosphere" (atm, which is now defined as 101.325 kPa or 760 mmHg at 0°C or 32°F) are still often used (Table 9).

The second method involves the use of an instrument that can calculate its position from a constellation of satellites (Section 4). Such devices provide the observer with three-dimensional coordinated, but their software is geared toward providing as much accuracy in the horizontal plane as possible at the expense of vertical accuracy. The resulting error in the elevation often appears greater than the error of a correctly used barometric altimeter. Some instruments combine both methods resulting in a more reliable reading, provided the barometric altimeter has been set correctly. In general it is best at a new site to establish the elevation of a single new bench mark, assume that the initial measurements or calculations are accurate, and accept any minor errors until a more accurate value becomes available and all previous results can be amended.

A map or plan is the projection of a three-dimensional reality, usually at a fixed reduced scale, on a two-dimensional plane, almost invariably defined as a perfectly horizontal. This plane can be a sheet of paper or digital screen. Obviously, most actual points of interest will be below or above this plane and the third dimension, or any other geospatial data, has to be added by using pre-defined symbols. One of the most practical and common methods is the use of contour lines (isopleths), which

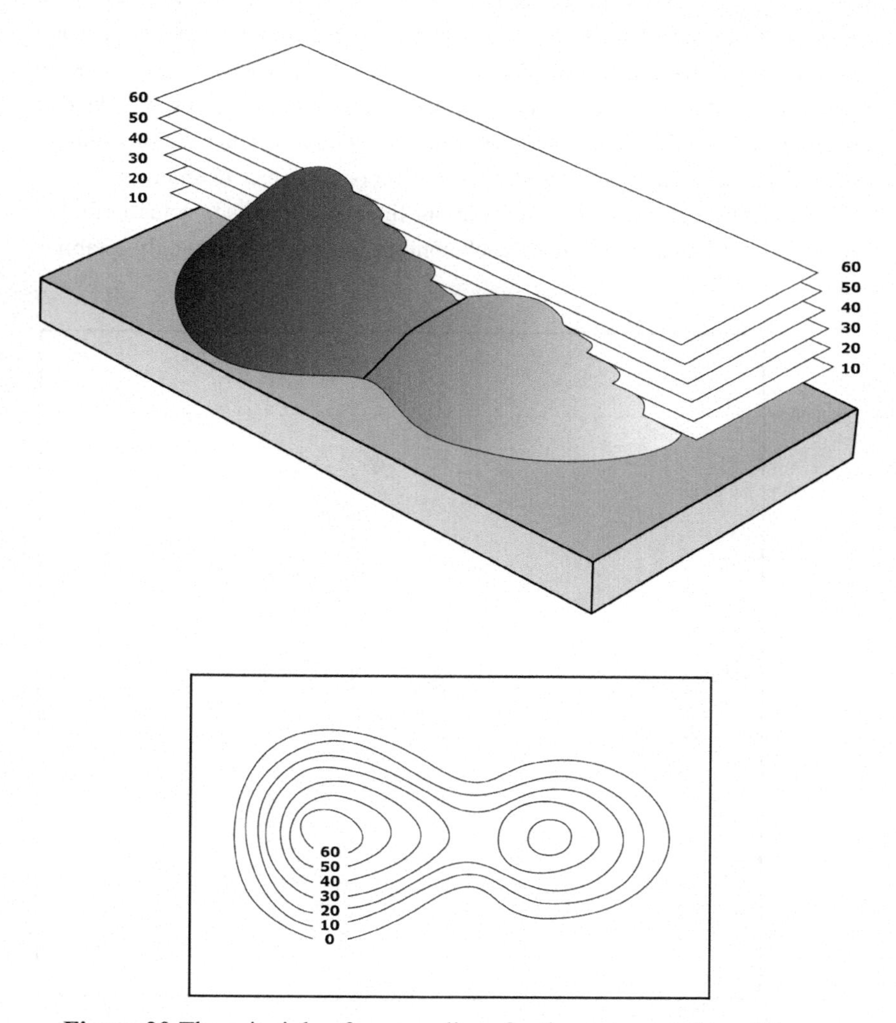

Figure 20 The principle of contour lines for the representation of three-dimensional topography on a two-dimensional map. Contour lines connect points with the same elevation (altitude), usually regular intervals apart

are lines that connect points with the same value of some continuous property, in the case of elevation the z-coordinate (Figure 20, Figure 22, top). Contour lines were originally developed at the end of the seventeenth century by Edmund Halley (1656–1742) to report on his research into the magnetic variation across the Atlantic Ocean (Section 3). Whereas Halley connected points in which the compass deviated from true north by the same angle, his method was quickly adopted to indicate the elevation of the terrain. For this, points with the same elevation, usually a round number that is increased with fixed steps, are connected by smooth lines (Figure 21). The same method has since been used to represent a wide variety of properties that can be linked to a specific point, ranging from air pressure (isobars) to the salinity of the ocean (isohalines), and from the cost of travel (isodapanes) to the frequency of the occurrence of the northern lights (isochasms). A special kind of contour map is a shaded contour map or heat-map (Figure 22), in which continuous geospatial data is represented by a color gradient or grey scale (Section 6).

The most accurate way to create contour lines on a map is to actually find in the terrain a number of points on each contour line and subsequently establish

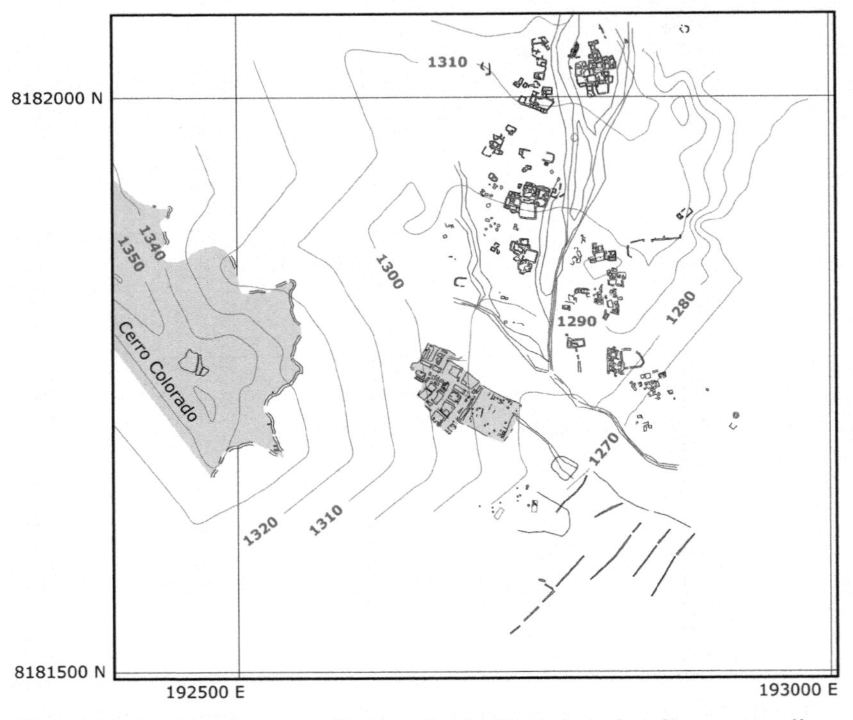

Figure 21 Section of a map of archaeological remains, showing contour lines (isopleths) connecting points with the same elevation above sea level

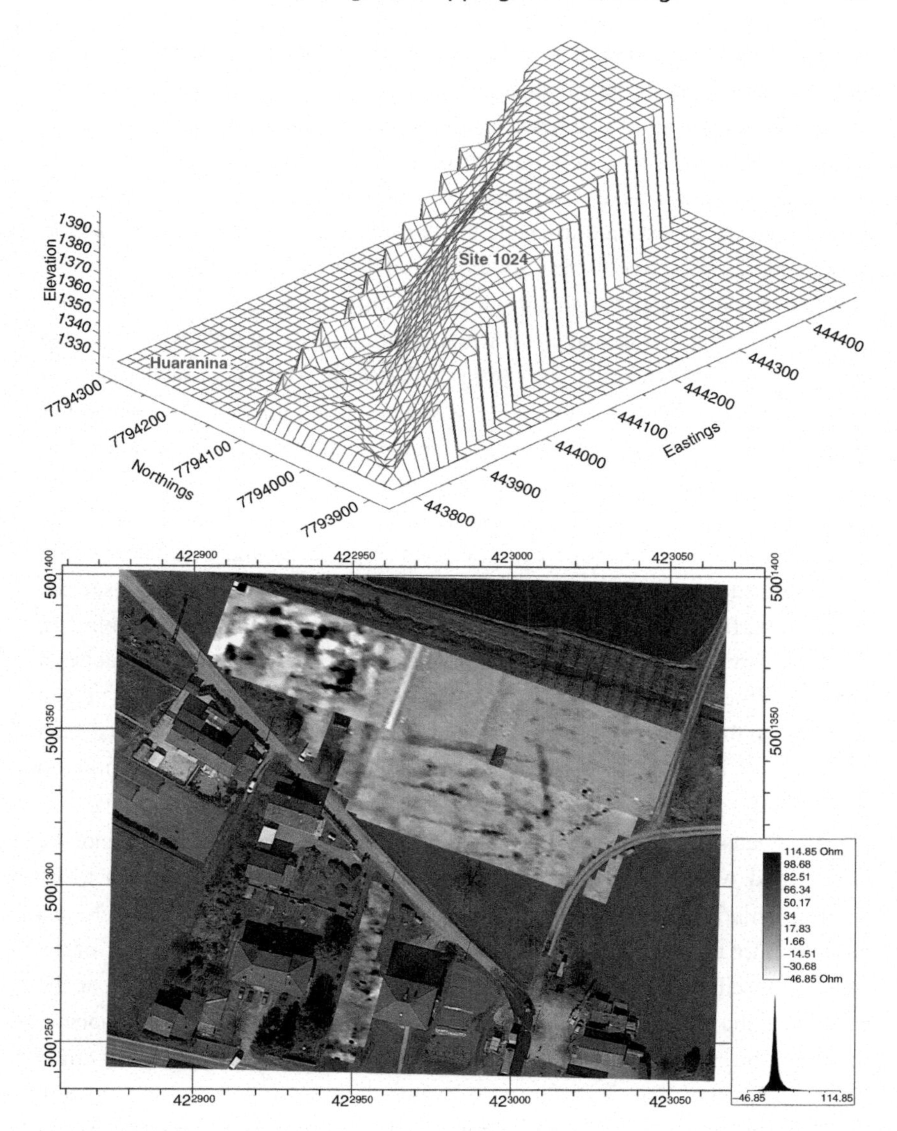

Figure 22 Elevational model showing the spatial relationship between two archaeological sites (top), see Figure 20. Heat-map indicating the electrical resistivity across an archaeological site inserted into a satellite image of the area (bottom)

their horizontal (x, y) coordinates. Each point can then be plotted and a smooth line can be drawn to connect points with the same z-coordinate. A second way to create contour lines is to establish the (x, y, z) coordinates of a large number of points that are more or less equally distributed across the landscape and label each with their elevation. Points of equal elevation can now be estimated and smooth contour lines can be draw in between the points of which the coordinates were actually measured. There are a number of software packages available that are able to do so automatically and accurately. Often the points of archaeological interest that were measured to create the two-dimensional map are sufficient to create contour lines, or only few additional points need to be added.

When working over longer distances, not only the topography of the landscape has to be taken into account when creating maps and plans but also the three-dimensional, spherical curvature of the earth. The shape of the earth is irregular, but can be approximated by a sphere or an ellipsoid. An even better approximation is the geoid, which is a regular mathematical body defined to resemble as closely as possible the shape of the earth at average ocean level, while ignoring local irregularities. Some of the properties of the earth can be demonstrated using the properties of circles (Figures 23 and 24). The straight line connecting two points that are a distance of 100 km apart along the surface of the earth, for instance, appears less than 100 m shorter (Figure 23, Table 10). This line obviously has to travel below the surface and remarkably reaches a depth of almost 250 m. The same properties show that for an average person standing on the beach the horizon is about 4.5 km away, while tall objects, such as towers or the mast of a ship, can sometimes be seen beyond the horizon (Figure 24).

Coordinates of points on the surface of the earth can obviously not be represented within a Cartesian grid and instead a spherical grid has to be created. This is usually done with the help of great circles. A great circle is a circle cutting through a sphere with its center coinciding with the center of the sphere. The circumference of a great circle is the largest that can be drawn on the surface of the sphere. The shortest distance between two points on a sphere, running along the surface of that sphere, is always a section of a great circle (Figure 23, Table 10). Great circles of which the circumference contains both the north and south poles are called meridians (Figures 25 and 26). The axis of the earth is a diameter of these great circles, while all meridians meet at both the north and the south poles. The great circle of which the circumference is equidistant from both poles is called the equator. Circles running parallel to the equator are called parallels. Apart from the equator no other parallels are great circles as they are getting increasingly smaller toward the poles.

Coordinates within the grid constituted by these meridians and parallels are usually expressed in degrees of latitude and longitude. The latitude of a point on the surface of the earth is the angle between the line going down from that point to the center of the earth and the line coming up from the center of the

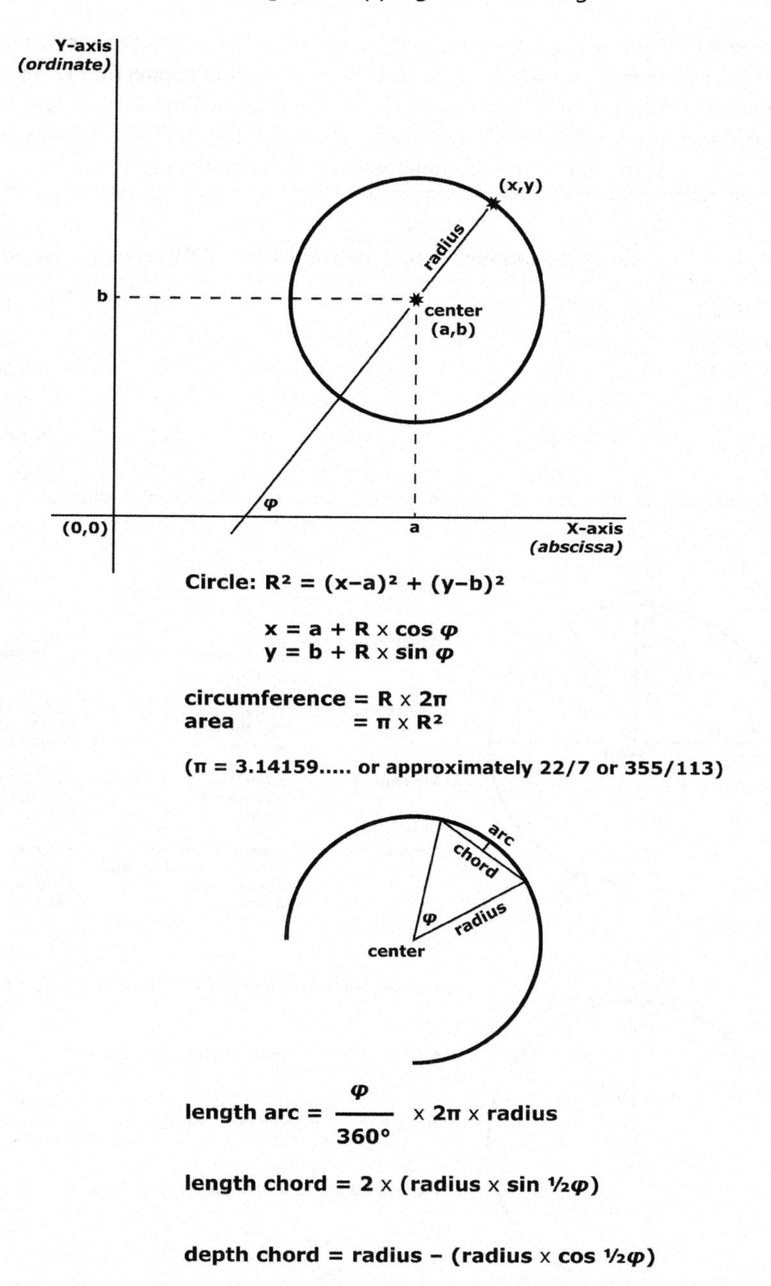

$$\text{length arc} = \frac{\varphi}{360^\circ} \times 2\pi \times \text{radius}$$

$$\text{length chord} = 2 \times (\text{radius} \times \sin \tfrac{1}{2}\varphi)$$

$$\text{depth chord} = \text{radius} - (\text{radius} \times \cos \tfrac{1}{2}\varphi)$$

Figure 23 General properties of a circle (top) and the length of arcs and chords (bottom), see Table 10

Table 10 Lengths of selected arcs and chords associated with the surface of the earth, see Figures 23 and 23. Note that the straight line connecting two points that are a distance of 100 km apart along the surface of the earth is less than 100 m shorter, while reaching a depth of almost 250 m. These figures are provided here to complement the text and illustrations

Angle	Distance along the surface	Distance in a straight line	Difference	Depth
15′ (0.25°)	27,808	27,784	24	15
30′ (0.5°)	55,615	55,568	47	61
45′ (0.75°)	83,423	83,352	71	136
1° (60′)	111,230	111,135	95	242
5°	556,150	555,506	644	6,061
10°	1,112,300	1,109,955	2,345	24,231

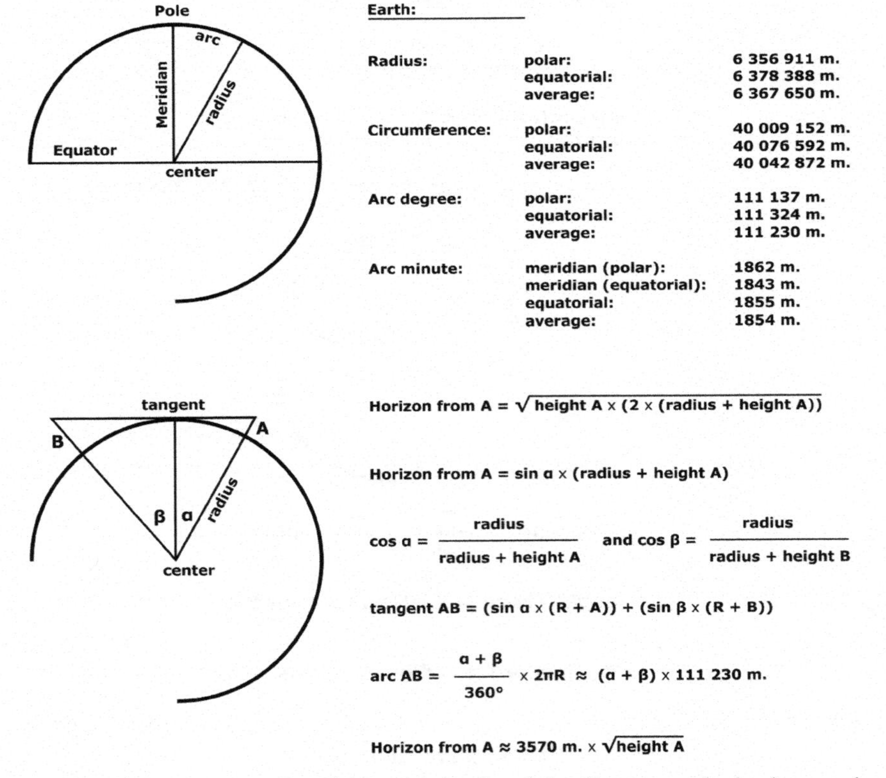

Figure 24 Some properties of the earth (top) and the distance to the horizon and objects visible beyond that (bottom). For an average person standing on the beach the horizon is about 4.5 km away

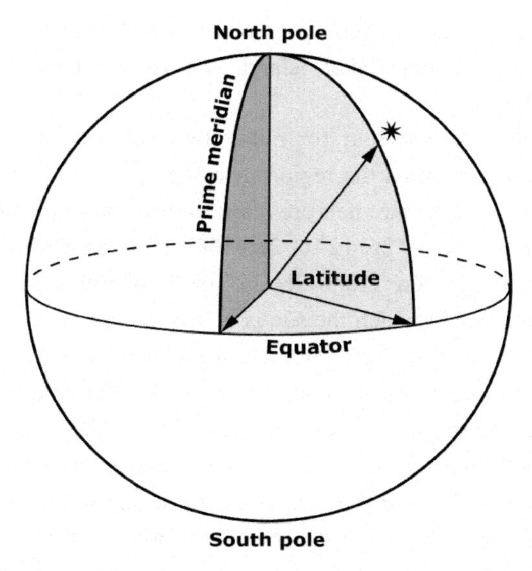

Figure 25 Latitude and longitude are expressed as angles between the point of interest on the surface of the earth, the center of the earth, and lines going up to the equator (latitude) and the prime meridian (longitude), respectively. This allows for the shortest distance (along the great circle) between two points, as well as the angle of the great circle with the meridian (azimuth) to be calculated, see Figure 26

earth to meet the equator (Figure 25). The longitude of a point is the angle between the line going down to the center of the earth and the line coming up to meet the prime meridian. The equator (0° latitude) is obviously fixed, being equidistant from the poles, but the prime meridian can be any of an infinite number of meridians. By modern convention, the prime meridian (0° longitude) is the meridian running across the observatory at Greenwich, near London, but in the past many different prime meridians have been used or proposed.

The prime meridian divides the earth in a western and an eastern hemisphere, with Asia, Australia, and most of Europe in the eastern hemisphere and the Americas in the western hemisphere. The equator divides the earth in a northern and a southern hemisphere, with Europe and Asia in the northern hemisphere and Australia and most of South America in the southern hemisphere. The prime meridian crosses the equator just west of Africa, near the island São Tomé, resulting in this continent being represented on each of the four hemispheres (the prime meridian and equator meet again on the other side of the earth, in the Pacific Ocean between Arorae Island, one the islands of the Republic of Kiribati, and Baker Island, one of the United States Minor Outlying Islands).

Apart from the equator, four (or rather two pairs) other parallels are of interest. The first pair limits the region in which the sun is directly overhead at least once a year. These parallels are referred to as the tropic of Cancer in the northern hemisphere (23°N 26′ 12″ or 23.43669°N) and the tropic of Capricorn in the southern hemisphere (23°S 26′ 12″ or 23.43669°S). The second pair limits the two regions in which the sun is below the horizon for a full day (24 hours) at least once a year. These parallels are referred to as the arctic circle on the northern hemisphere (66°N 33′ 48″ or 66.56331°N) and the antarctic circle on the southern hemisphere (66°S 33′ 48″ or 66.56331°S). These angles correspond with the tilt of the axis of the earth respective to its orbit around the sun. In the temperate regions between the tropics and the arctic and antarctic circles, the sun is never perpendicular to the surface of the earth and remains visible every day of the year.

The advantage of this system of representing positions on the surface of the earth, and the main reason it was developed, is that is allows for the distance along the great circle, and thus the shortest distance possible, between two points with known spherical coordinates – latitude and longitude – can be calculated (Figure 26). This is obviously of great importance to travelers, especially those traveling long distances in a ship or an airplane. It also allows for the angle to be calculated at which the great circle crosses the meridians. This angle changes continuously as one travels along the great circle. To avoid frequent complicated calculations Gerardus Mercator (Geert de Kremer, 1512–94) suggested the well-known and still frequently used Mercator projection. Mercator projected the spherical earth on a cylinder, which was subsequently rolled out into a flat map. On this map all meridians are parallel and cross the parallels at right angles (Figure 27). A straight line on such a map thus crosses all meridians at the same angle, greatly simplifying navigation, which can now be accomplished with only a compass (Section 3). The sacrifice made by traveling a longer distance, because the apparently straight line is not a great circle and thus not the shortest distance between the two points, was considered more than worthwhile in times when arriving at the right place was much more important than the time traveled. The Mercator project distorts the outlines and surface area of countries and continents, and dramatically so farther from the equator, but that too was considered a minor disadvantage.

Another advantage of the system expressing coordinates on the surface of the earth as latitude and longitude is that it allows for the calculation of

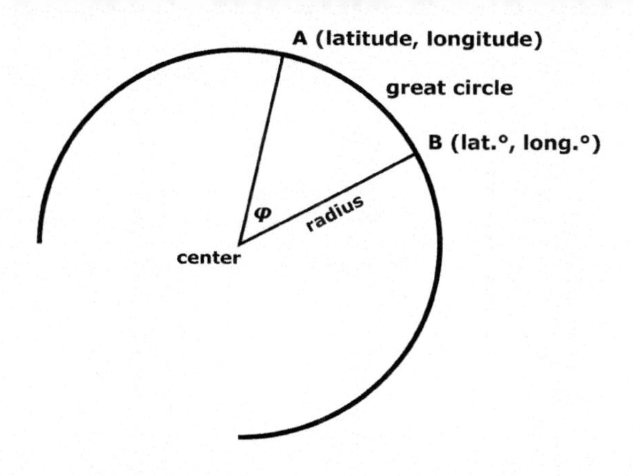

$$\cos \varphi = (\sin \text{lat.A} \times \sin \text{lat.B}) + (\cos \text{lat.A} \times \cos \text{lat.B} \times \cos(\text{long.A} - \text{long.B}))$$

$$\sin \alpha = \frac{\cos \text{lat.B} \times \sin(\text{long.A} - \text{long.B})}{\sin \varphi} \quad \text{(azimuth)}$$

$$d = \frac{\alpha}{360^\circ} \times 2\pi R = \alpha \times 111{,}230 \text{ m.} \quad \text{(distance)}$$

Figure 26 The formulae to calculate the distance and azimuth between two points with known coordinates (latitude and longitude) on the surface of the earth, see Figure 25. Electronic survey equipment and software will have these equations embedded

the coordinates of unknown points by celestial observations. This requires an instrument to accurately measure angles, such as a sextant or a survey instrument; knowledge of the position of heavenly bodies, such as the sun, the planets or selected stars; and the exact time. This technique is not discussed here as it has almost entirely been replaced by devices that calculate their position from the signal of a constellation of satellites (Section 4).

3 Practical Mapping and Planning: Finding North

The English word orientation is rooted in the word orient (east), referring to the religious desire to find the direction of the rising sun or the city of Jerusalem, the center of early Christianity. The practice to pray and construct sacred spaces facing a specific direction has mostly faded from Christianity, but is still prevalent in, for instance, Judaism and Islam. Modern maps are no longer oriented east, but rather with north at the top. There are, however, three slightly different types of north. First and probably best known is magnetic north,

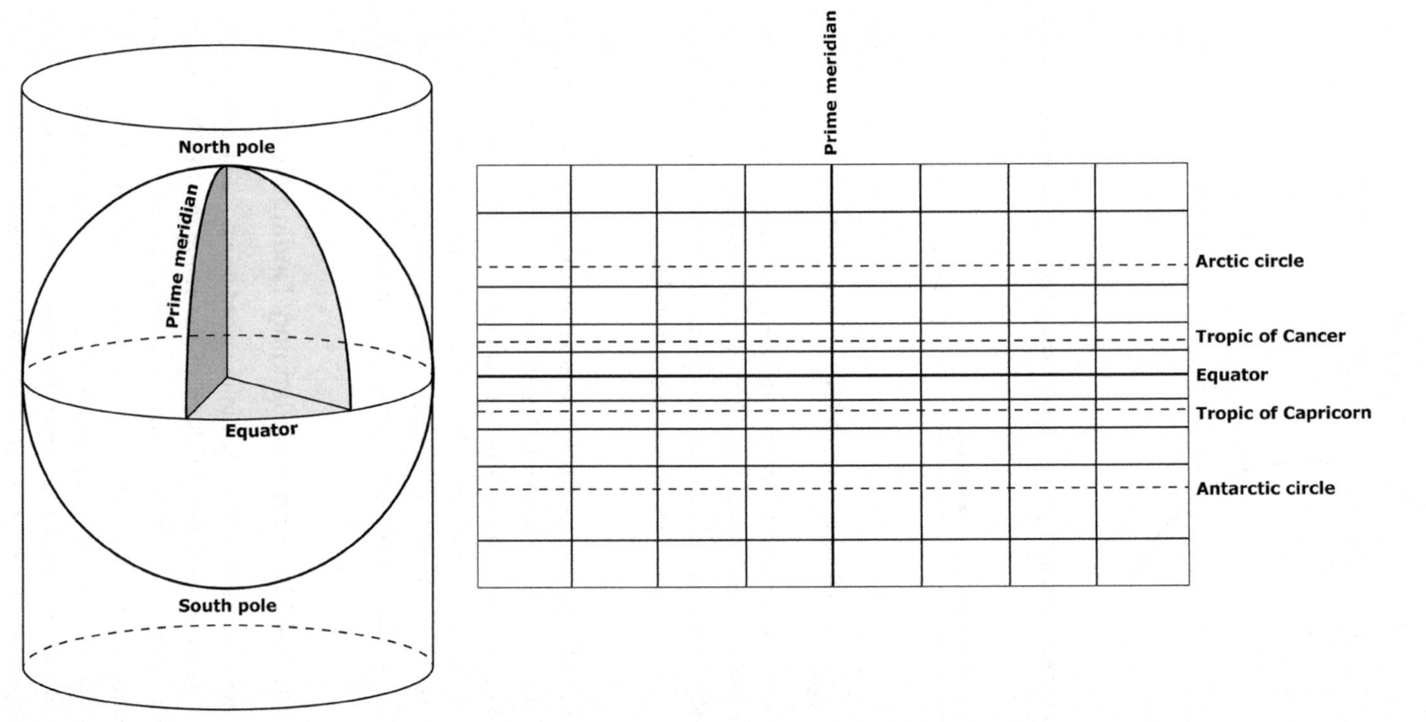

Figure 27 The Mercator projection converts the three-dimensional surface of the earth into a cylinder, which can be rolled out into a two-dimensional map, see Figure 41. On this map both parallels and meridians are parallel and cross at right angles, while surface areas are progressively distorted north and south of the equator

or rather geomagnetic north. This is the direction indicated by a compass, when used in the right conditions. Second is grid north (geodesic or map north), which is north along the grid lines of the projection employed to draw the used map. Third is geographic north (astronomical or true north), which is the direction toward the point where all meridians cross at the north pole. Good topographic maps will have all three directions indicated, with a date accompanying the direction of magnetic north as this slowly changes over time.

A compass is a magnetic needle that is suspended in a way that allows it to move freely along its vertical axis. Usually the needle is in a disk-shaped container with its movement dampened by a liquid filling the container. Given that the earth has a relatively strong magnetic field, for reasons not entirely understood, the needle of a compass will arrange itself parallel to the magnetic field of the earth and point toward its magnetic north and south poles. As these more or less coincide with the geographic north and south poles, albeit not exactly, compasses have been used to indicate north for centuries. As the needle will locally always point in the same direction, compass can also be used to create simple maps.

Traditionally, the horizon is divided into four equal parts in which four cardinal directions are recognized: north, east, south, and west. Halfway between these are four ordinal or intercardinal directions: northeast (NE), southeast (SE), southwest (SW), and northwest (NW). Dividing each section into halves again results in a 16-point compass rose (Figure 28). Many different systems have been suggested and employed to further subdivide this circle, but for the sake of mapping and planning a division into 360° is the most

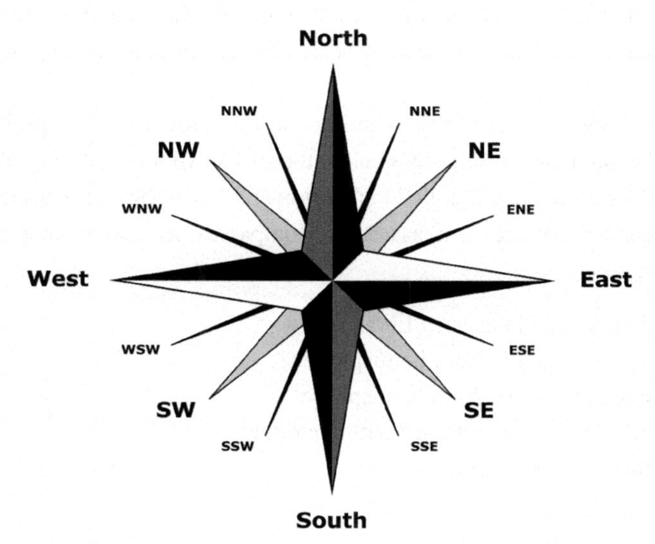

Figure 28 The 16-point compass rose, see Figure 4

appropriate and convenient. Conventionally, north corresponds with 0°, east with 90°, south with 180°, and west with 270° or –90° (Figure 4).

When establishing magnetic north with a compass, three important issues need to be taken into consideration: declination, inclination, and deviation. The declination is the local difference between magnetic north and true (geographic) north. As the magnetic north and south poles do not accurately coincide with the geographic north and south poles, there is a difference between the direction in which the compass needle points and geographic north and south. If the observer, the magnetic north pole, and the geographic north pole line up, this difference will be zero. East and west of that position the difference will increase until it reaches its maximum 90° away, measured along the parallel, after which it will again decrease to zero, after traveling 180° along the parallel. In principle, it should thus be possible to establish longitude by measuring the difference between true north and magnetic north. This idea was investigated in the seventeenth century by Edmund Halley, who established the magnetic variation across the Atlantic Ocean and produced a map on which he connected points in which the compass deviates from true north by the same angle (Figure 29), one of the first known examples of representing geospatial data by contour lines (isopleths). There are, however, also significant local influences, such as topography and variation in the composition of the crust of the earth, that render the method impractical. The magnetic field of the earth is furthermore not stable but slowly drifts (Figure 30) and apparently at times completely reverses, when the magnetic north pole moves to a position near the geographic south pole, and vice versa (Figure 31). The latter seems to take place at irregular intervals and the time frame remains unknown; it may happen overnight or take weeks, months, or years.

The tip of the compass needle pointing north is at the same time pulled down because the magnetic north pole is below the surface of the earth (Figure 23 and Table 10). This tendency is called inclination and will to be compensated for by the manufacturer of the compass. Most compasses are made for one of five inclination zones:

1. North America and northern Eurasia.
2. The tropics on the northern hemisphere.
3. The tropics on the southern hemisphere.
4 Southern Africa and southern South America.
5. Australia and New Zealand.

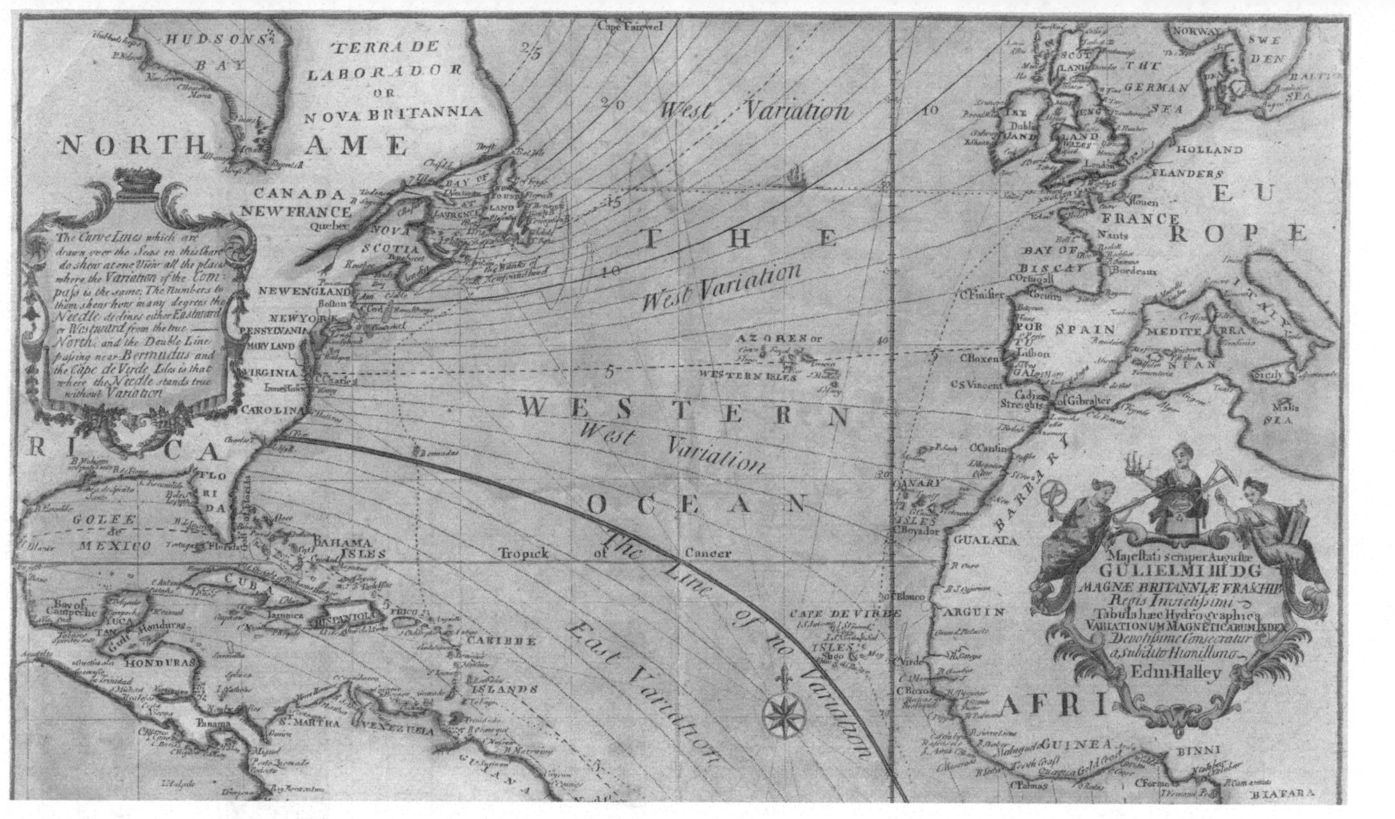

Figure 29 Section of the map created by Edmund Halley, around 1700, showing lines connecting points with the same variation of the compass across the northern Atlantic Ocean. This is one of the first known examples of contour lines (isopleths) to visualize geospatial data. Note that the variation can be as large as 15–20°. Image courtesy of Princeton University Library, Special Collections, Historic Map Division (Firestone Library HMC01.6613)

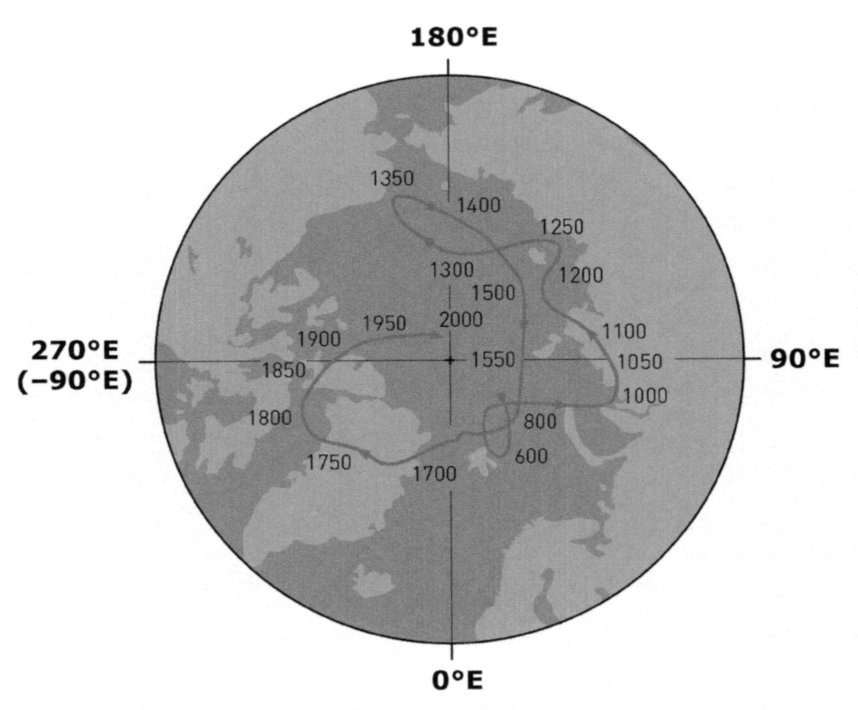

Figure 30 The position of the magnetic north pole between around 500 and 2000 CE

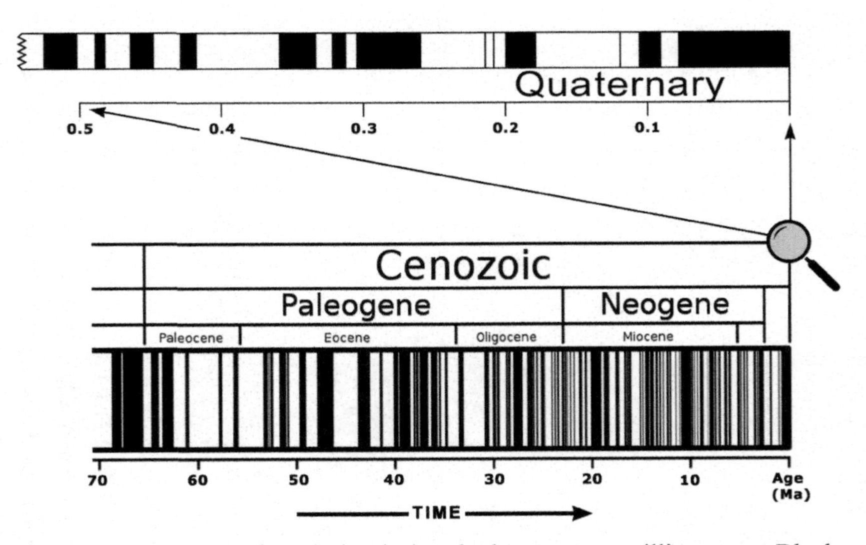

Figure 31 Geomagnetic polarity during the last seventy million years. Black bands indicate time periods during which polarity was it occurs today (at the right), while white bands indicate the reverse (with the magnetic north pole near the geographic south pole). The last reversal occurred around 780,000 years ago, making the current polarity last for a relatively long period

The point where the needle of the compass would point straight down is referred to as the north magnetic pole, which is close to yet different from both the geomagnetic north pole and the geographic (true) north pole.

Some compasses can be adjusted to function properly in different inclination zones. Others can be adjusted to read the correct azimuth in different declination zones, and some can be adjusted for both. The third external influence on the compass are errors that are the result of very localized sources of magnetism such as magnets, ferrous metals, and electrical equipment. Such errors should be avoided whenever possible by moving away from their source. Compasses that are fixed in place, for instance in ships and aircrafts, are usually compensated for the unavoidable but stable disturbing influences in their direct surroundings.

Despite these issues, compasses have the useful property that they will locally always point in the same direction, albeit not necessarily exactly north. In premodern times, this enabled sailors to navigate their ship after losing sight of land by steering at a continuous angle with the needle of the compass. Maps on which the spherical surface of the earth is projected on a cylinder, the Mercator projection (Figure 27), enable this angle to be established by drawing a straight line between the place of departure and the destination. This is not the shortest distance between the two – which would be the great circle which crosses every meridian at a different angle (Figure 26) – but it allows for simple and secure navigation. For centuries such certainty was more important than distance or time of travel.

The same property allows for a compass to be used as a basic survey instrument, similar to a level instrument (Figures 16 and 17). Compasses made for this purpose, sometimes referred to as geological compasses, have a simple or more complex sight and can usually be mounted on a light tripod (Figures 32 and 33). Using the direction in which the compass needle points as a base line, oriented more or less north–south, polar coordinates can be established (Figure 6). Each point of interest can be located by the combination of its distance from the observer, which can be scaled down when the points are plotted on a map, and the angle of the line of sight with the base line, which remains unchanged at any scale. As obvious from the legacy of John Gardner Wilkinson (Section 2), it is also useful for any surveyor to know the length of her or his stride, allowing distances to be estimated fairly confidently by counting paces. If such becomes necessary or more convenient, the compass can be moved to another position, which is first located and plotted, where it will point in the same direction, facilitating an easy continuation of the survey activities. This method enables relatively large areas to be mapped and to move around optical and physical obstructions such as standing buildings, vegetation, or topography.

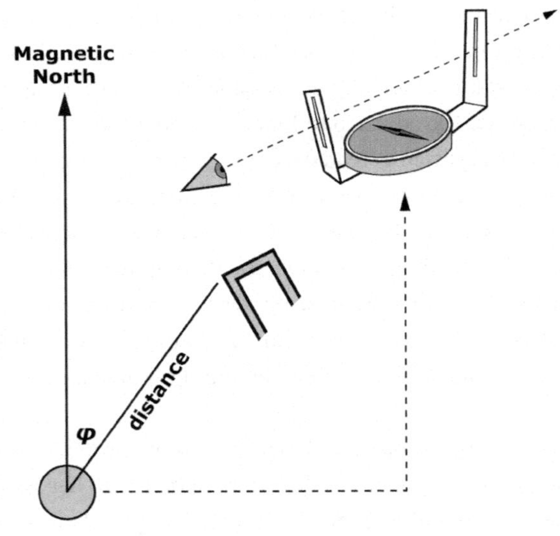

Figure 32 With a sighting (or geological) compass, in combination with a distance measuring device, simple maps can be created by establishing polar coordinates of points of interest, see Figures 6 and 33

Figure 33 An archaeologist establishes polar coordinates for ancient remains with the aid of a sighting (geological) compass and a measuring tape, see Figure 32

A compass can thus not be relied upon the indicated true north, but there are other and more accurate methods to find north. Traditionally, these are based on celestial observations, hence the synonym astronomical north. Every day the sun passes the meridian, and thus due south (on most of the northern hemisphere) or due north (on most of the southern hemisphere) of the observer. This happens when the sun reaches its highest position (zenith) that day, almost exactly halfway between sunrise and sunset. The moment that this happens and the direction toward the sun at the time can be relatively easily established, especially in the temperate zones. If the horizon is more or less clearly visible, the angle between the observer and the place of sunrise and sunset can be recorded, preferably on the same day. The line dividing this angle in half will run along the meridian (Figure 34, left). This can be done by simply placing a number of markers (lengths of wood or metal) in the terrain and dividing the angle between the center of observation and the direction of the sunrise and the sunset with a survey instrument or a measuring tape. With sufficient knowledge of the sky at night, this method can be applied to every celestial body that is not circumpolar (always above the horizon).

It is important to keep in mind that the telescope of even a simple survey instrument will focus the energy of the sun, which can immediately and irreversibly damage the retina and cause permanent loss of vision. Solar filters are available for survey instruments and no survey instrument should ever be used for solar observations without such a filter. Even looking into the sun with the naked eye for a prolonged period of time can permanently damage the retina.

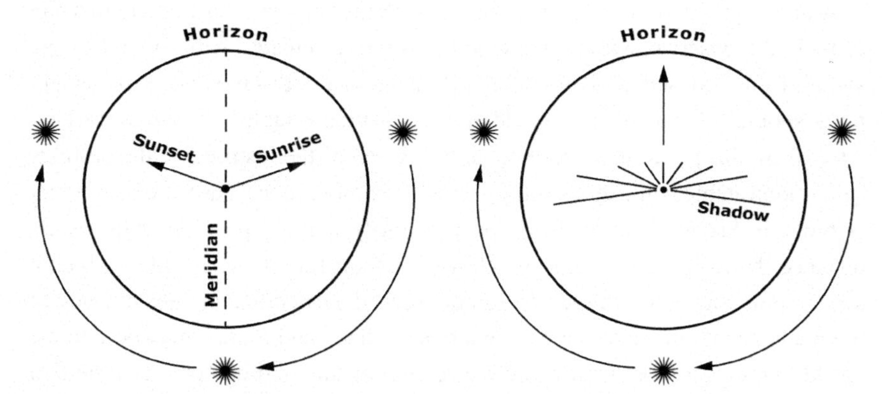

Figure 34 Finding the meridian by observing the movement of the sun. *Never look into the sun with a survey instrument or for a prolonged period of time.*

Never look into the sun with a survey instrument or for a prolonged period of time

Another relatively easy and accurate method to find north using the movement of the sun across the meridian is to observe the shadow of a stick placed vertically into the ground, effectively functioning as a gnomon. Around noon, the tip of the shadow can be marked and measured at regular intervals, with the shortest shadow pointing exactly north on most of the northern and exactly south on most of the southern hemisphere (Figure 34, right). Toward the equator, the angles between sunrise and sunset, as well as the angle of the sun when it reaches its zenith, change unfavorably, rendering these methods increasingly difficult to use across the tropics. Furthermore, it has to be kept in mind that in the tropics the sun can be either north or south of the observer, depending on the time of year. In the period March–September, the sun can be north of the observer in the northern hemisphere tropics and in the period September–March, south of the observer in the southern hemisphere tropics.

Two other methods enable an accurate determination of the direction of true north. The first can be employed in the northern temporal zone, between the tropic of Cancer and the arctic circle, where the relevant stars are visible and can be easily observed with a survey instrument. Polaris, the northern star, is always within 1° off true north, providing a quick and relatively accurate method to find north. Using the vertical crosshair in the telescope, however, it is possible to wait for Ursa Maior ζ (zeta) or *Mizar* (Arabic for belt), Polaris, and Cassiopeia δ (delta) or *Rukbah* (Arabic for knee) to line up, which is when Polaris crosses the meridian (Figure 35). This provides a very accurate vector toward geographic north, although only available in the northern hemisphere.

A final method to establish true north is with the help of an instrument that can provide the observer with a geographic position calculated with the aid of the signals from a constellation of satellites (Section 4). For this method, a convenient point should be marked in the field and its position established. When walking away from this point, while looking at the screen of the instrument, the eastings can be kept stable while the northings slowly increase. In that case, the instrument is moving due north away from the previously marked position. After some distance the instrument should be allowed a short time to settle, after which a second point due north of the first can be marked. Depending on the accuracy of the readings and the distance traveled, this method can be used anywhere in the world to create a more or less precise section of the meridian between the two marked positions (Table 11). Some of these instruments are also outfitted with a magnetic compass, which is as reliable as the mechanical compass described earlier in this section. Other instruments calculate the direction of north from

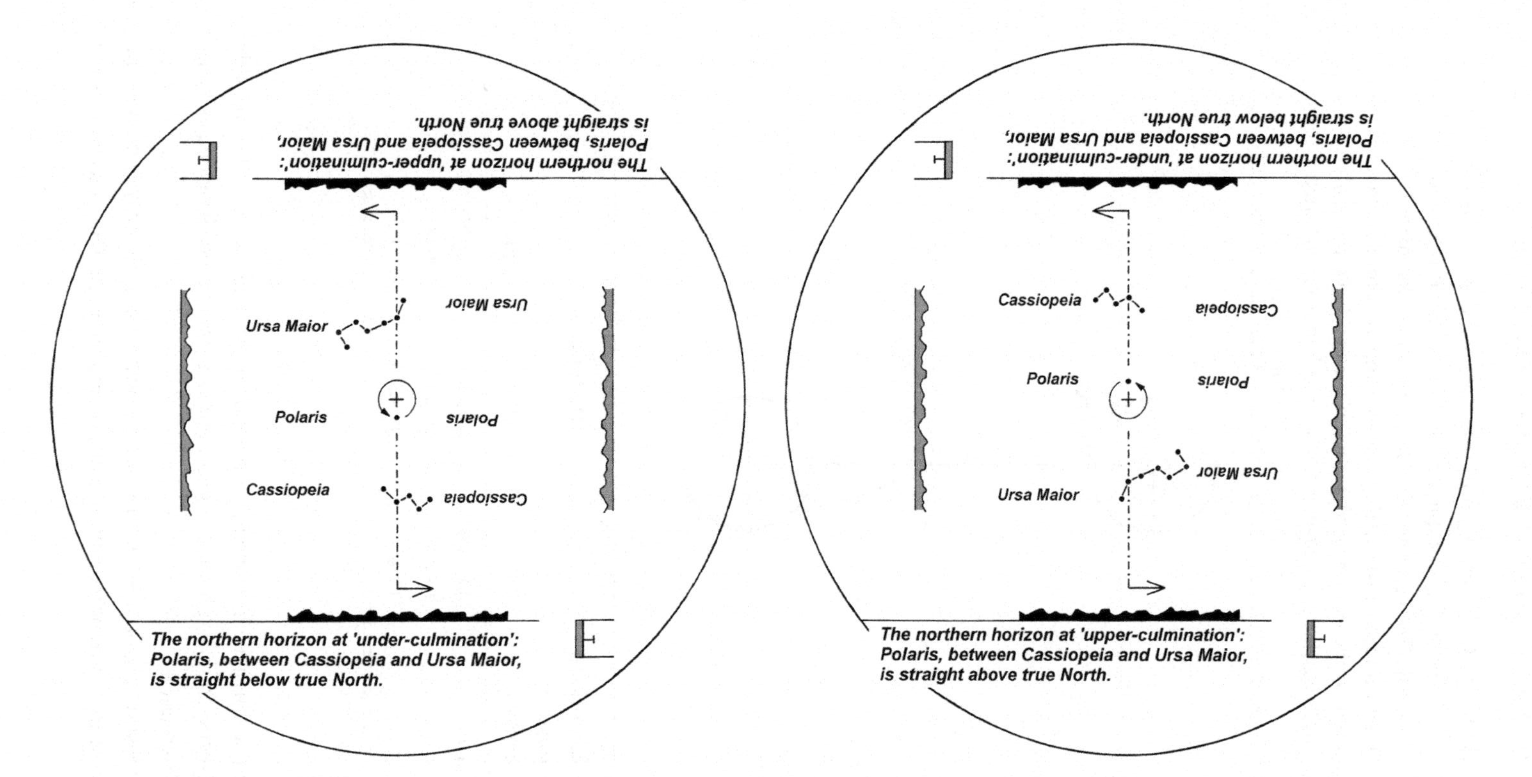

Figure 35 Finding the meridian by celestial observation of Polaris (the northern star), Cassiopeia, and Ursa Maior

Table 11 Maximum error off true north using a device that can provide a geographic position calculated from the signals from a constellation of satellites (estimated position error, provided by the instrument, and distance between the two located points in meter). Note that only high-resolution measurements yield better results than celestial observations

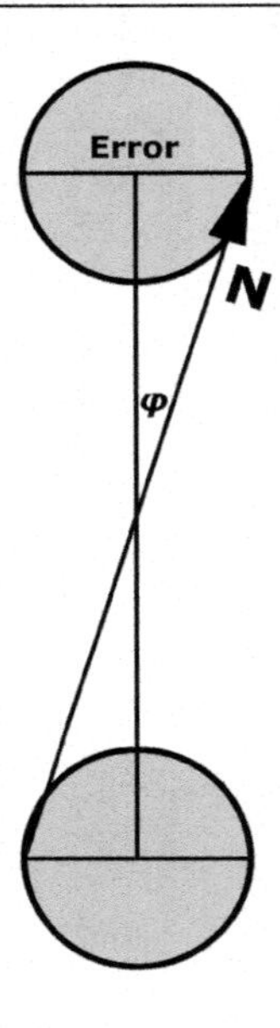

Estimated position error	Distance 50	100	200
10	11° 19′	5° 43′	2° 52′
5	5° 43′	2° 52′	1° 26′
1	1° 9′	0° 34′	0° 17′

readings taken when the instrument is in motion. In principle, this is more reliable than a magnetic compass, but less controlled than the methods based on celestial or solar observations.

It is not just magnetic north which slowly drifts, but also the direction of the axis of the earth – currently pointing toward Polaris (Figure 35) – as well as the

Table 12 The average tectonic movement of the major plates forming the outer crust of the earth

Plate	**Displacement (cm/year)**
Cocos	8.6
Pacific	8.1
Nazca	7.6
Philippine	6.4
Indian	6.0
Arabian	4.7
Caribbean	2.5
African	2.2
Antarctic	2.1
South American	1.5
North American	1.2
Eurasian	1.0
Average	4.3

plates that form the crust of the earth and carry the continents and oceans. Their average displacement is about 4 cm (1½ inch) per year, or a little less than half a meter in a decade (Table 12). This tectonic movement is the cause of earthquakes and volcanic eruptions, among many other geological phenomena, and ensures the constant renewal of our inorganic environment. This is one of the factors that enable life on earth, but at the same time it displaces carefully aligned ancient structures as well as our fixed points, bench marks, and survey monuments. For most archaeological projects, which rarely last much more than a decade, this is seldom a significant issue, but it is something to keep in mind when leaving markers for a next generation or studying the orientation of ancient structures.

4 Practical Mapping and Planning: Field Walking

The term "archaeological survey" can refer to the accurate mapping of structures, objects, and layers within an archaeological site or excavation unit, as well as to the more or less systematic identification and location of ancient remains in the landscape. The former is sometimes referred to as topographic survey, mapping, or planning, the latter as field walking. Field walking can be limited to general reconnaissance – identifying ancient remains in a relatively unstructured, yet time-efficient fashion – or entail a more systematic and time-consuming approach. Independent of the research design, the instrument of choice is a device that can calculate its position with the help of the signal from a constellation of satellites. The first and most important of these is the Global Positioning System (GPS), launched and maintained by the United States, which lends its name to the instruments. Other, newer constellations include the Chinese BeiDou Navigation Satellite System, the Russian Global Navigation Satellite System (GLONASS), and the Galileo Global Navigation Satellite System, launched and maintained by the European Union. A list of the most often used abbreviations is provided in Table 16.

To find their position, GPS instruments have to receive the signal broadcasted by at least three satellites of the same constellation and thus need to be able to see the sky, or at least a sufficiently large section thereof. Once connected, they will receive an almanac which contains precise information on the geographic position of each satellite in the constellation. The only additional information needed is the distance from the instrument to each of the satellites within view. Combined with the presumption that the observer is somewhere on the surface of the earth, this allows the instrument to calculate its position, within a certain margin of error, the estimated position error (EPE). The accuracy of the instrument is limited by the very short times required for the signal to travel between the satellites and the instrument, the speed of the satellites relative to the observer, atmospheric disturbances, the tidal movements of the surface of the earth, and relativistic effects. The best accuracy that a simple handheld instrument is able to reach is 3–5 m (Figure 36, top). When drawn on a map the position should be reflected as a circle with a diameter of the EPE provided by the instrument. For most large-scale archaeological survey 3–5 m is sufficient and acceptable. The algorithms used, however, aim to return the best possible horizontal accuracy (x, y), at the expense of vertical accuracy (elevation or altitude). Elevations provided by any GPS instrument should therefore not readily be accepted (Section 2). Given their low cost, durability, and ease of operations, handheld GPS instruments have become a standard tool among archaeologists (Figure 37).

The accuracy of this method of locating geographic positions can be greatly increased by one of several methods referred to as differential GPS (DGPS). For this method, a second receiver is placed in a known position, or allowed to establish its position by averaging a large number of readings (10,000–1,000,000). Once the actual geographic position is entered into or calculated by the stationary instrument, it subsequently establishes the difference between the position calculated from the signals of the satellites and its known position, usually at 1 second intervals. Many of such base stations have been placed around the world. Some of these broadcast the difference that they measure back to the satellites, where a second signal is created that contains the average

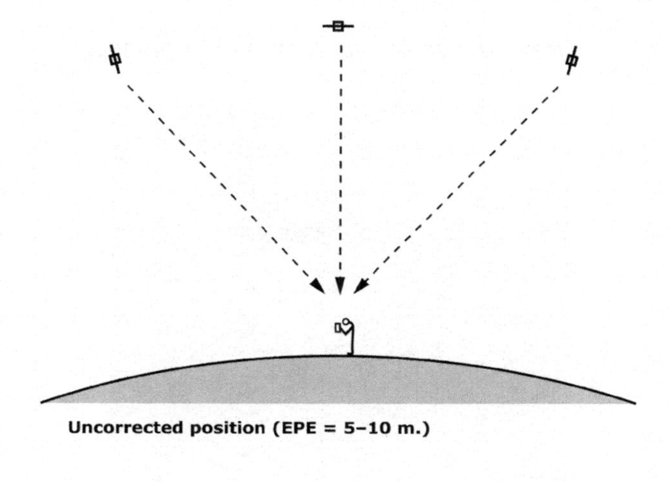

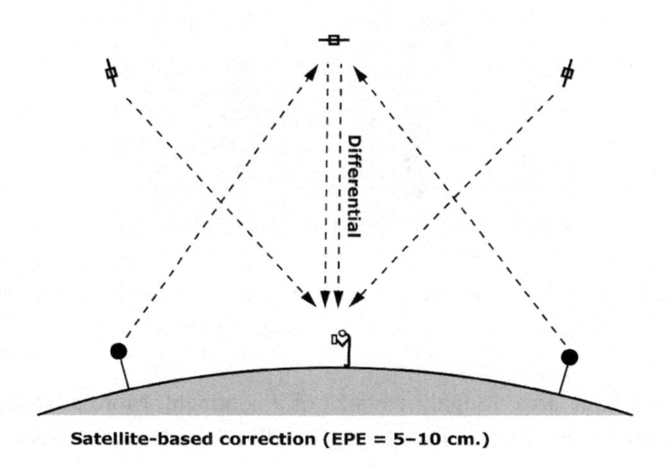

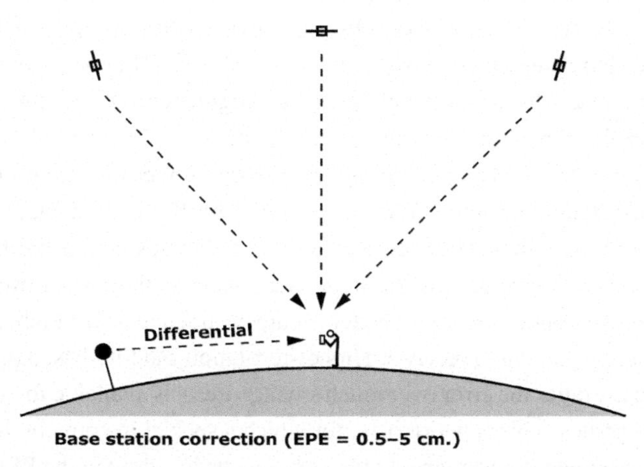

Figure 36 Schematic overview of the various methods to establish a geographic position using the signal from a constellation of satellites; see Figures 37, 38, and 39

Figure 37 A handheld GPS device used to collect archaeological geospatial data with an EPE of 5–10 m; see Figure 36 (top)

error over a large area, roughly the size of a continent. More advanced GPS receivers are enabled to connect to this satellite-based augmentation system (SBAS), resulting in a drop of the EPE to around 1 m (in the horizontal plane). The most important SBASs include, among others, the European Geostationary Navigation Overlay Service (EGNOS), operated by the European Union; the GPS Aided Geo Augmented Navigation system (GAGAN), operated the Indian government; the Multi-functional Satellite Augmentation System (MSAS), operated by the Japanese government; the System for Differential Correction and Monitoring (SDCM), operated by the Russian government; and the Wide Area Augmentation System (WAAS), operated by the United States Federal Aviation Administration. The accuracy of this combined signal is limited by the fact that errors are averaged over a large area, whereas they will differ locally. There are several options to attain better accuracy, albeit at a cost higher than the prize of a simple handheld receiver. Higher-resolution satellite-based differential data, which averages the error over much smaller areas, is available for a fee from several companies. This superior data, for which a special instrument that is able to receive the signal is necessary (Figure 38), results in a drop in the EPE to 5–10 cm (in the horizontal plane), about ten times better than a handheld receiver alone.

Figure 38 A differential GPS (DGPS) device used to collect archaeological geospatial data with real-time correction resulting in an EPE of 5–10 cm; see Figure 36 (middle). Photographs courtesy of John Steinberg and Willeke Wendrich

For even better accuracy access has to be gained to a local base station, preferably not farther than 50–100 km away. These are often present in light houses, airport control towers, and government and academic buildings. Some broadcast their differential data for all nearby to use, others are linked to selected instruments. Receiving such high-resolution differential data also requires a special instrument. As archaeological projects are often in remote areas with limited infrastructure, it may be necessary to purchase and install a dedicated base station (Figure 39). Some of these come with a data logger which collects differential data throughout the work day, while low-resolution data is collected in the field, with an instrument referred to as a rover (Figure 40). As each reading, both by the rover and by the base station, comes with an accurate time stamp, the data collected in the field can later be corrected by applying the differential correction calculated by the base station.

The disadvantage of this process, called post-processing, is that no high-resolution data is available in the field. Other base stations broadcast their differential signal to the rover, allowing for real-time correction. Connecting to a local base station can result in an EPE of 0.5–5 cm, depending on the distance to the base station (Figure 36, bottom), about a hundred times better than a handheld receiver alone. This is sufficient for more complex survey work to be done, such as the detailed planning of individual structures. One method to address the issue with the vertical accuracy, which is inherent to all instruments although it increases with better horizontal accuracy, is to locate one or more known points during each survey. This allows for all collected elevations to be corrected and the survey to be conducted with internal consistency. This also renders it possible to amend the measured topography of the whole site once more accurate positions of these fixed points become available. To facilitate this it is helpful to locate any monument left by others and try to retrieve the

Figure 39 A dedicated stationary base station used to collect differential GPS data which allows archaeological geospatial data collected by a rover to be post-processed to an EPE of 0.5–5 cm; see Figures 36 (bottom), 40, and 66

Figure 40 A rover used to collect low-resolution GPS data on archaeological structures, which can be post-processed to an EPE of 0.5–5 cm with the data collected by a nearby stationary base station; see Figures 35 (bottom), 39, and 66. Photographs courtesy of Brett Kaufman and Danny Zborover

coordinates that they assigned to these (Figure 41). To locate is used here to mean to establish the coordinates of a specific point in three-dimensional space.

The geographic location established by GPS instruments is usually returned as longitude and latitude (Section 2), expressed in degrees–minutes–seconds or decimal degrees (Section 1). The reason for this is that it allows for the shortest distance – along the great circle – between points with known spherical coordinates to be calculated as well as the azimuth, the angle of the great circle with the meridian (Figures 25 and 26). At the scale that most archaeological work takes place, however, a major disadvantage of this system is that it is difficult to visualize, while the mathematics are tedious (Section 5). For this reason, most archaeological surveyors use the Universal Transverse Mercator (UTM) system, developed during the Second World War by both the German and

Figure 41 Government monuments (survey markers permanently fixed in place) found in Chile, Italy, Tunisia, and the United States (clock-wise from top-left)

American armed forces to address the same issues encountered by archaeologists. The UTM system envisions the world as divided into sixty north–south strips, each 6° of longitude wide (Figure 42). Instead of curved, rounded lozenges with sharp points at the poles (Figure 43), these strips are represented as flat rectangles onto which the surface of the earth is projected. This dramatically distorts distances and surface areas, especially farther from the equator, but allows for coordinates to be represented as Cartesian coordinates (Section 1). Such coordinates can easily be visualized and enable distances between points with known coordinates to be calculated using the Pythagorean theorem (Figure 8). This is, however, only possible within a zone and does not work for two or more points in different zones.

Within each zone the coordinates of a point are represented by Zone–Easting–Northing (ZEN). The zone is simply the number of the zone (1–60) in which the point is located. Originally, each zone was divided into twenty bands (C–X, skipping O and I to avoid confusion), which is written right after the number of the zone. Los Angeles (California), for instance, is in zone 11S, and Lima (Peru) in zone 18L. The number of bands is sometimes reduced to two: N for the northern hemisphere and S for the southern hemisphere. As both

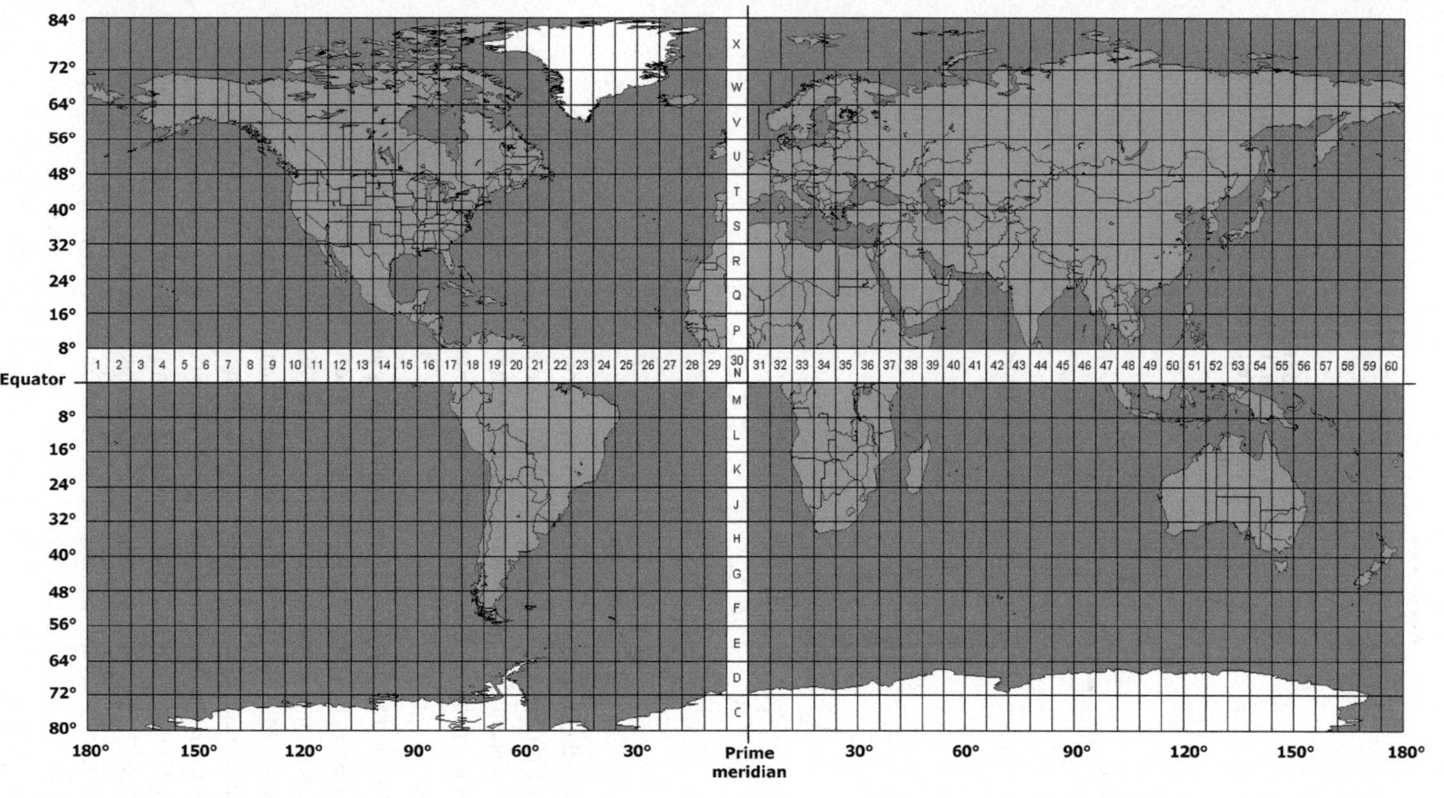

Figure 42 Map of the world in Universal Transverse Mercator (UTM) projection, indicating the sixty UTM zones (labeled 1–60, from west to east) and the twenty UTM bands (labeled C–X, from south to north, skipping I and O to avoid confusion); see Figure 27. Confusingly, sometimes only two bands are used: N (northern hemisphere) and S (southern hemisphere)

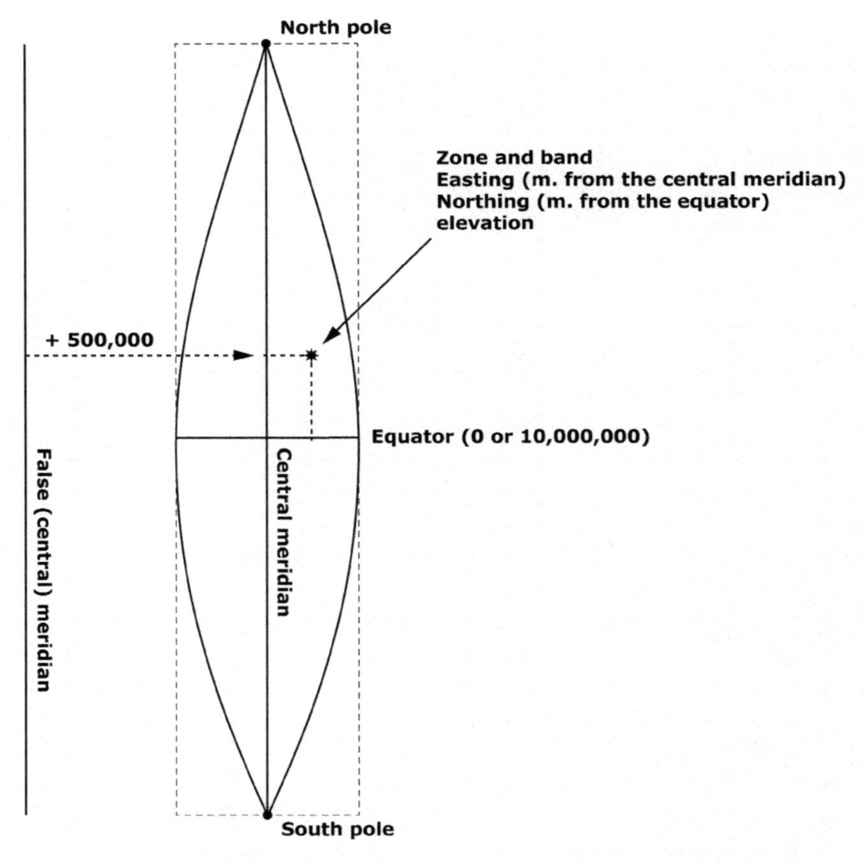

Figure 43 Schematic representation of UTM coordinates as Zone–Easting–Northing (ZEN). Easting = distance (in m) from the central meridian + 500,000. Northing = distance (in m) from the equator on the northern hemisphere; and 10,000,000 – distance from the equator (in m) on the southern hemisphere

N and S also signify bands in the more detailed system, this simplification can lead to confusion. It is therefore better to write either "11 North" and "18 South" for Los Angeles and Lima, respectively, or "11S (N)" and "18L (S)."

Within each zone the position of every point can now simply be expressed in Cartesian coordinates, meaning meters from the central meridian (easting) and meters from the equator (northings). If known, the elevation (altitude) is given after that. If used like this, however, each number might appear four times, in both the eastings and the northings and both positive and negative (Figure 5). Furthermore, on the northern hemisphere the northings would increase traveling north, while on the southern hemisphere they would increase traveling south. To avoid possible confusion an arbitrary 500,000 m is therefore added to all eastings, so that eastings are always positive and increase going east. Additionally, on the southern hemisphere the distance to the equator is subtracted from an arbitrary 10,000,000 m to make all northings positive and always increase going north (Figure 43). Applying these rules, the position of Los Angeles (34°N 05′–118°W 15′) is represented as 11S (N)–384600 E–3768400 N in the UTM system, while the position of Lima (12°S 05′–77°W 00′) is 18L (S)–279000 E–8667100 N (or simply 11S 384600 3768400 and 18L 279000 8667100, respectively).

When using a GPS device several additional issues have to be considered, apart from the basic units (metric or miles, see Section 1), the desired north (geographic or magnetic, see Section 3), and the coordinate system in which positions are to be displayed (longitude–latitude or UTM, as discussed earlier in this section). These choices can easily be entered into all instruments, while most instruments store the raw data allowing for conversions to be made after the data has been collected. When calculating positions from the signals provided by the satellites points are not projected onto a flat horizontal plane nor on the actual surface of the earth, but rather on the less undulating, yet curved surface of a theoretical ellipsoid (Section 2). This ellipsoid is chosen to coincide as much as possible with the actual surface of the earth, but will obviously cut through higher regions and pass over low-laying areas. Until recently most national survey authorities used an ellipsoid that was locally very accurate, but possibly relatively far off elsewhere in the world. Many of the 1:50,000 and 1:25,000 topographic maps used in archaeology are projected on such local ellipsoids, and using another ellipsoid may result in points appearing as much as 100–200 m off. The used ellipsoid is always given in the legend or key to the map and the instrument to be used should be set to the same ellipsoid for the data to fall in the right place. Some historic ellipsoids that are often encountered in archaeological work include Airy 1830 (Great Britain), Clarke 1866 (United States), Clarke 1880 (Africa), Helmert 1906 (Egypt), and South American 1969. Most instruments can project their readings on a range of ellipsoids, while instruments that store the raw data allow for conversions to be made later. Transcontinental aviation and other elements of globalization created a need for an ellipsoid that could be used throughout the world. In 1972, the World Geodetic System (WGS72) ellipsoid was accepted, which was further refined in 1984 (WGS84). Most new maps are projected on this WGS84 ellipsoid, although many older maps remain in circulation. For this reason, or reasons of national pride, some countries persist on using ellipsoids other than the worldwide standard.

Elevations can be calculated in meters above the ellipsoid, in meters above sea level, or in meters above a local vertical datum or bench mark. Many countries have established their own bench mark with an elevation more relevant to the local situation than an elevation above sea level or any ellipsoid. This too is indicated on the topographic maps used. For archaeological work any of these bench marks can be used, as long as the choice is clear and consistent. Some of such local bench marks include the North American Datum 1983 (NAD83) and the Provisional South American 1956 (PSA56) datum.

Next to its position and altitude, a GPS instrument provides additional information which can be more or less relevant for the ongoing survey work. All instruments will show which constellation is used to calculate the given position, including the number of satellites in view, the quality of the signal of each, and the number of satellites actually used for the calculation of the position. If the satellite geometry becomes unfavorable or the most useful signals weaken, the displayed EPE will increase. In this case it is advisable to pause the survey work for 10–15 min and wait for the satellites to move into better positions. If the instrument is in a steady motion, it can easily calculate its speed and heading from successive positions, as well as the direction of geographic north (Section 3). Most instruments, however, also contain a magnetic compass and will display magnetic north if the instrument is stationary. This can differ significantly from geographic north although some instruments contain information on the local declination. If accurate orientation is necessary, however, it is better to do so in a systematic fashion (Table 11). In any case it is important to calibrate the internal compass at least daily, or at the start of every survey activity, in the manner indicated by the instrument (usually by rotating it a full circle along each of its three axes).

Other instruments contain a barometric altimeter which can be used to improve the altitude calculated by the instrument (Figure 19, Table 9). For this the altimeter needs to be calibrated regularly at a point with a known elevation and be moved as swiftly as possible to the points with unknown positions (Section 2). All instruments can tell time very accurately, which is at the basis of the calculations, and most can also provide the local times of the rising and setting of the sun and the moon. Many instruments furthermore have more or less detailed maps built in, which at the very least allow a quick visual check of the data. A final useful function of most instruments is the ability to find previously located points (waypoints). When activated, this function displays the distance and direction from the current position of the observer to the desired destination. By simply following the directions provided by the instrument, this allows the observer to find previously recorded locations within the given margin of error.

Finding and locating ancient remains spread across an area can be done following three different methods: opportunistic, full-coverage survey, and sampling. It is important to decide in advance which method is going to be followed, as they require different research layouts that will prove difficult to merge once started. First, the research area (universe) needs to be defined and a location system established (Figure 44, top). This can be the geodetic grid (longitude–latitude or UTM) or a grid system created to fit the local situation

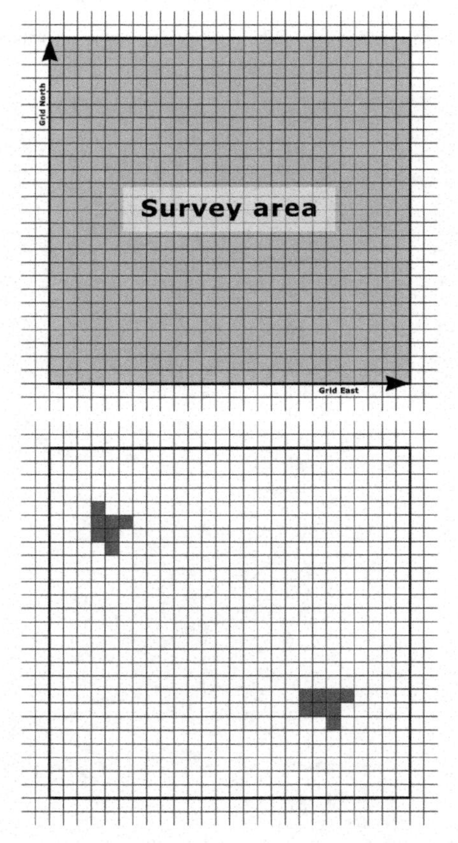

Figure 44 Archaeological survey should take place in a clearly defined area or universe, with a system to locate objects or structures of archaeological interest (top). Marked squares indicate ancient remains (bottom). This schematic map can be the result of either opportunistic or full-coverage survey activities. Areas can usually be identified by the remains found on the surface (dwelling, kiln, food, or tool production site, etc.)

(for instance 10 m^2 parallel to the edges of the research area). Within this confined and controlled area ancient remains can be found and located by traveling across the landscape, preferably on foot, but alternatively, if circumstances so dictate, by car or on horseback (Figure 45, top).

The chances of finding ancient remains can be greatly increased by starting near known ancient sites, following known ancient routes, but especially by asking local farmers – who obviously have profound knowledge of their farmlands – or interviewing otherwise knowledgeable local informants. During any survey it is of greatest importance to find rapport with the local population and gain their trust and understanding. This will greatly increase the yield of the survey work and at the same time prevent the disturbance or destruction of ancient remains. This opportunistic approach is especially

Figure 45 Archaeologists perform an extensive, opportunistic survey (top) and intensive, full-coverage field walking (bottom)

suitable for larger areas, kilometers rather than meters, and in regions where archaeological remains are readily visible in the landscape (Figure 44, bottom). Although they provide valuable information, the results of such an opportunistic survey are not suitable for statistical analysis.

Smaller and more significant or complex areas can be investigated with a full-coverage survey. For this method a group of archaeologists line up about 2 m apart and systematically walk up and down the survey area until every part of it has been checked for archaeological remains (Figure 45, bottom). Usually, survey flags or other markers are planted at each find (Figure 46, top), which are later located with a GPS instrument or otherwise (Figure 47). A full-coverage survey is obviously the most secure, but at the same time the most

Figure 46 Archaeologists place and locate flags marking archaeological finds (top); see Figure 47, and sample an ancient site using the dog-leash method (bottom); see Figures 48 (bottom) and 49

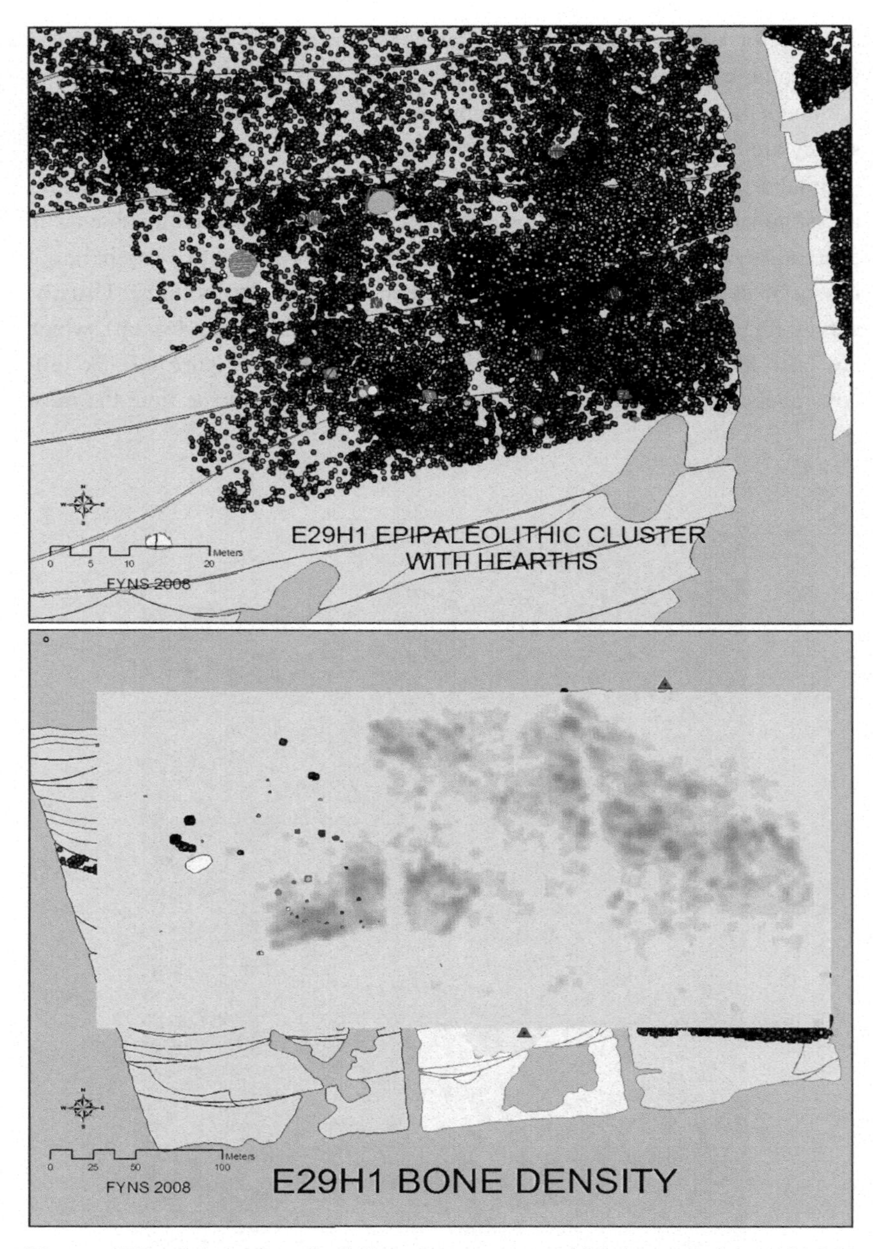

Figure 47 Partial results of a full surface survey, which located about 100,000 points where stone flakes or processed bone were found (top). Heat-map based on the full surface survey represented directly above, indicating the relative density of processed bone. Images courtesy of Simon Holdaway and the Fayum Project

time-consuming method to record the archaeological potential of a specific area (Figure 44, bottom).

At times an area of interest is densely covered with archaeological remains, yet too large for a full-coverage survey to be deployed, or at least within the available time frame and budget. In such a case, a sample of the area can be investigated and the assumption made that the sample represents the area as a whole. For this to be valid the distribution of the places to be investigated across the survey area has to be random, as any nonrandom pattern will introduce structure into the data set and will render it unsuitable for statistical analysis (Figure 48, top). There exist a variety of methods to generate a random sample of a large area. These are beyond the scope of this Element, and only a few often used methods will be briefly discussed here to provide an overview of the possibilities.

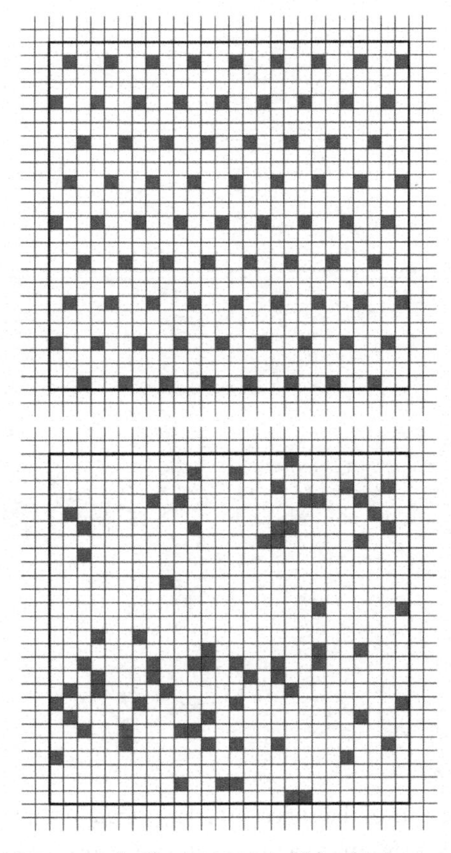

Figure 48 Marked squares indicate areas within the larger survey area to be sampled. The pattern at the top is not random and should always be avoided as it will introduce structure into the data set. The pattern at the bottom shows the same number of marked squares, but randomly distributed across the survey area, allowing for statistical analysis of the data

The squares within the survey area can be numbered and a random subset of these selected with the aid of a printed table or generated by dedicated software or a website (Figure 48, bottom). Instead of square boxes, areas can also be investigated with the so-called dog-leash method (Figure 46, bottom). For this a

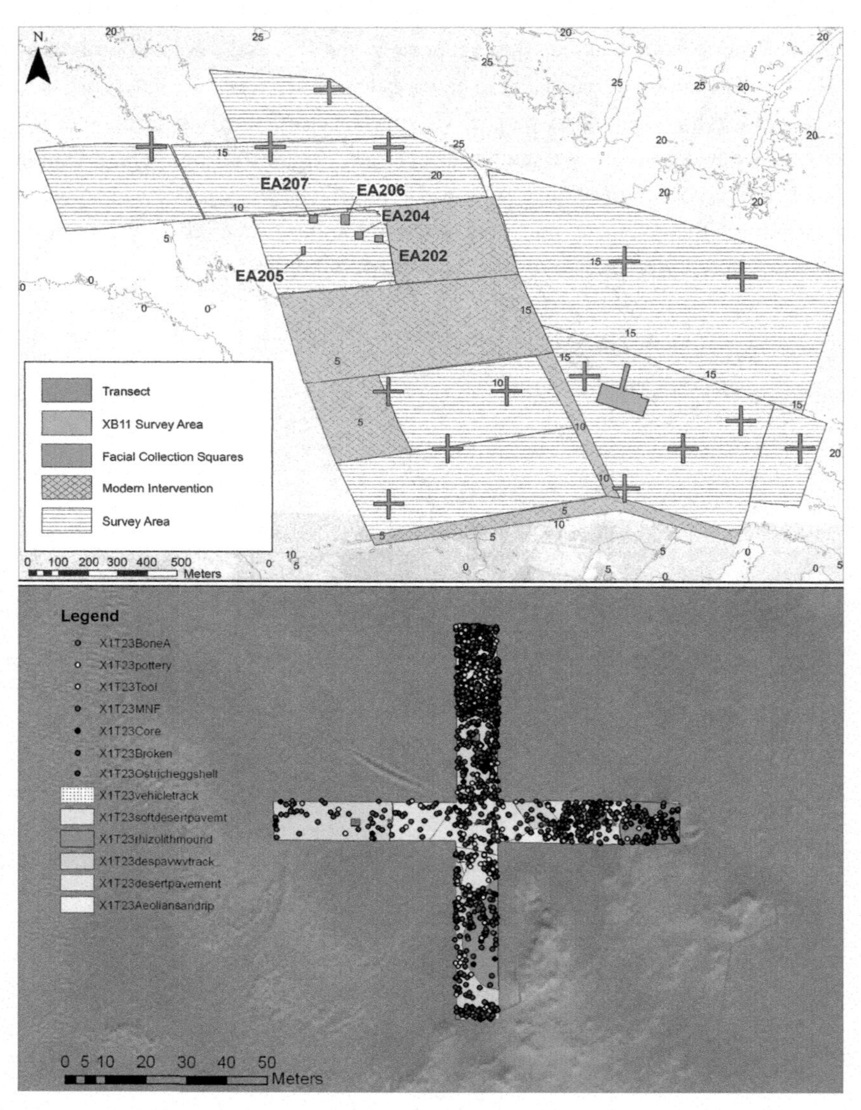

Figure 49 Partial results of the survey of a large area of archaeological interest, sampled using both a stratified sampling strategy (top) and a statistical model which determined the best position of the next survey area, consisting of two crossing strips of 10 × 100 m each (bottom). Images courtesy of Simon Holdaway and the Fayum Project

number of random positions within the survey area is generated. At each of these positions a pin is placed to which a string with a known length is attached (usually 5 m). All archaeological materials within the circle created by the string, with the pin at its center, are subsequently located, studied, and if necessary collected. Areas where archaeological remains are poorly visible can similarly be subjected to a so-called shovel-test (or bucket-test). For this a small hole is opened at each of the random positions across the survey area and either a shovel or a bucket full of sediment is examined for archaeological artifacts. Provided care has been given to make the sample random, such methods can reliably establish the archaeological potential of a specific area.

More sophisticated methods to make a randomized surface survey more productive and informative include stratified sampling strategies, and sampling strategies based on statistical modeling. For a stratified sampling strategy the survey universe is divided into several sub-regions, based on criteria such as the characteristics of the surface, the nature of the present vegetation, the topography, and so on. A specifically designed sample of each specific region is subsequently collected. A variety of this is to create a statistical model that can determine the best position of the next sample area based on the results of the previously studied areas (Figure 49).

5 Practical Mapping and Planning: Measured Plans and Maps

As any archaeological excavation irretrievably destroys the archaeological record while it is being recovered, it is of vital importance to record as much of the inferred information as soon and as best as possible. Such recording includes ample photographs and notes, but also measured plans of the layers as they are being excavated as well as the profiles left in place in sections, or baulks, to show unexcavated layers in the vertical plane. Exposed architecture is also recorded in detail, in the horizontal as well as vertical plane, and archaeologically significant objects are often recorded in situ, before being removed for additional study. Measured plans can be created from rectified photographs or digital measurements taken with electronic survey instrument, but complimentary to these are (partial) hand-drawn maps and plans. Each of these recording techniques, and more, do not render another dispensable, rather they are complementary. Among the reasons that drawings are still a valuable medium is that they allow for interpretations to be reflected – by highlighting or instead disregarding specific details – and, more importantly, that they provide an opportunity to study the ancient remains and their intricate relationships to the extent necessary to produce an accurate drawing. The most important skill that an archaeological surveyor should develop, or have, is the ability to visualize a mental image of the area to be planned and imagine what the final map should look like. Making a sketch map before starting the actual survey work will not prove just helpful in this, but near indispensable. Additional useful skills include a constant awareness of the cardinal directions and the length of one's stride. Digital methods and technologies are complementary to these basic skills and cannot fully replace them as an effective avenue to gain insights in exposed ancient remains (Section 1).

Excavation units are almost invariably squares or rectangles (Table 13), with four right angles (Figure 50, Video 2), and their sides can thus readily be used as the axes of a local grid system (Section 1). Coordinates can be established either directly (Figures 6 and 51, top) – aided by two or more tape measures or a planning frame (Figure 51, bottom) – or by triangulation (Figures 7 and 52, top), and transferred to scale to a drawing (Table 3, Figures 51 and 52). For vertical surfaces to be drawn (Section 2), such as profiles or standing walls (Figure 52, bottom), a line level and a plumb bob can be used to create two axes at a right angle in a vertical plane (Figures 14 and 53). Measurements taken with a tape measure should always be parallel to the plane onto which the measurements are projected or be amended to reflect the correct length (Tables 14 and 15). After

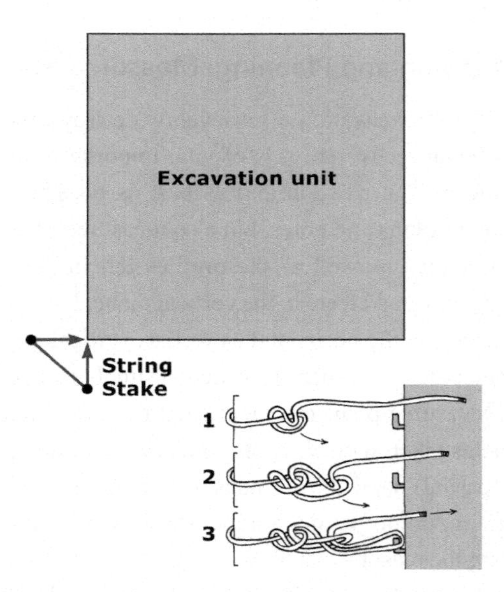

Figure 50 The most convenient way to string an excavation unit is to place two rather than one corner stakes a little away from the unit and have the strings cross at the actual corner (top). This ensures that the corner stakes will not get dislodged when the unit is excavated. Strings can be tightened with a so-called trucker's hitch (Ashley 1114), consisting of two successive slip knots (bottom)

Table 13 Length of the diagonals of selected squares and rectangles. Note the so-called Pythagorean triangles 3–4–5 and 6–8–10

	1	**2**	**3**	**4**	**5**	**6**	**7**	**8**	**9**	**10**
1	1.414	2.236	3.162	4.123	5.099	6.083	7.071	8.062	9.055	10.050
2	2.236	2.828	3.606	4.472	5.385	6.325	7.280	8.246	9.220	10.198
3	3.162	3.606	4.243	5	5.831	6.708	7.616	8.544	9.487	10.440
4	4.123	4.472	5	5.657	6.403	7.211	8.062	8.944	9.849	10.770
5	5.099	5.385	5.831	6.403	7.071	7.810	8.602	9.434	10.296	11.180
6	6.083	6.325	6.708	7.211	7.810	8.485	9.220	10	10.817	11.662
7	7.071	7.280	7.616	8.062	8.602	9.220	9.899	10.630	11.402	12.207
8	8.062	8.246	8.544	8.944	9.434	10	10.630	11.314	12.042	12.806
9	9.055	9.220	9.487	9.849	10.296	10.817	11.402	12.042	12.728	13.454
10	10.050	10.198	10.440	10.770	11.180	11.662	12.207	12.806	13.454	14.142

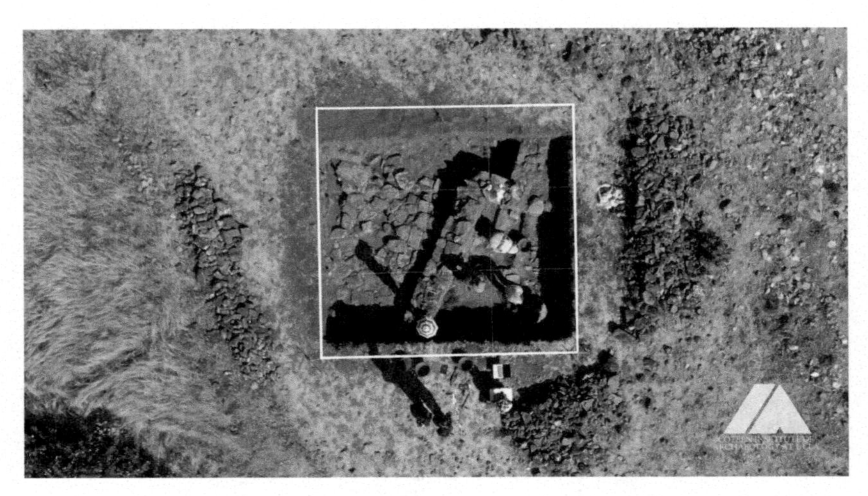

Video 2 Field archaeologists demonstrate how to create a right angle (90° or ½π) and a square excavation unit. Video available at www.cambridge.org/barnard

digitization, hand-drawn maps and plans can be augmented by photographs (Figure 54, top), or vice versa (Figure 54 bottom).

Larger-scale survey work, such as the planning of sites or surface, finds across a relatively small area is usually done by collecting polar coordinates of all relevant points and plotting these to scale in a digital environment. To collect polar coordinates two different instruments are needed, one to measure distances and a second to measure angles, both in the horizontal and the vertical planes (Figure 12). Some of these instruments are indeed sometimes still used, such as a level instrument combined with tacheometry (Section 2, Figures 16 and 17), a geological compass combined with a tape measure (Section 3, Figures 32 and 33), or a transit or theodolite combined with a tape measure. Currently, however, the instrument of choice is a total station, which combines these two types of instruments into one. As all measurements are done electronically, large amounts of raw data can be stored, which allows them to be collected swiftly and accurately, and subsequently stored, manipulated, and visualized.

A total station is a telescope mounted on a base (tribrach) in such a way that it can rotate smoothly along both its horizontal and vertical axes. These movements are very accurately measured, returning the angles of movement in both planes. As not all instruments are designed the same way, it is important to establish first which units are used, often D–M–S, but sometimes gons or decimal degrees (Section 1). Next, the plane of vertical measurements needs

Figure 51 Archaeologists record ancient remains on a measured plan with the aid of Cartesian coordinates; see Figures 7 and 63. Photographs courtesy of Willeke Wendrich

to be found, which can be done by rotating the telescope until the vertical angle reads 0°. This is often straight up, but can also be horizontal or straight down. No such fixed directions are available for the horizontal angle and all instruments will have a setting to define a specific direction of the telescope as 0° (grid north). The units for distances can usually be changed between the metric (meters) and the imperial or engineering (feet and inches) system (Table 2). Most instruments, however, will not return any readings unless first set up properly.

Like a level instrument, a total station should be either in its box or secure on a tripod in a perfectly level position. Unlike a level, however, a total station often also needs to be set up in a specific place from which the measurements

Figure 52 Archaeologists record ancient remains using triangulation (top), and Cartesian coordinates in a vertical plane (bottom). The axes are created by two strings: one kept horizontal with the aid of a line level, the second kept vertical with the aid of a plumb bob; see Figures 7, 14, 53, and 64

will be taken (Figures 6 and 12). There are several ways to achieve this, the most convenient of which is described here. To take correct readings, and often even to return readings at all, the survey instrument has to be perfectly level in a location with known, calculated or assigned coordinates, referred to as the occupied point. To accomplish this, the tripod is placed with its top roughly horizontal at a convenient height with the legs of the tripod more or less evenly spread out around the occupied point. A convenient height means that the eyepiece of the telescope is at the eye level of the person who will be using

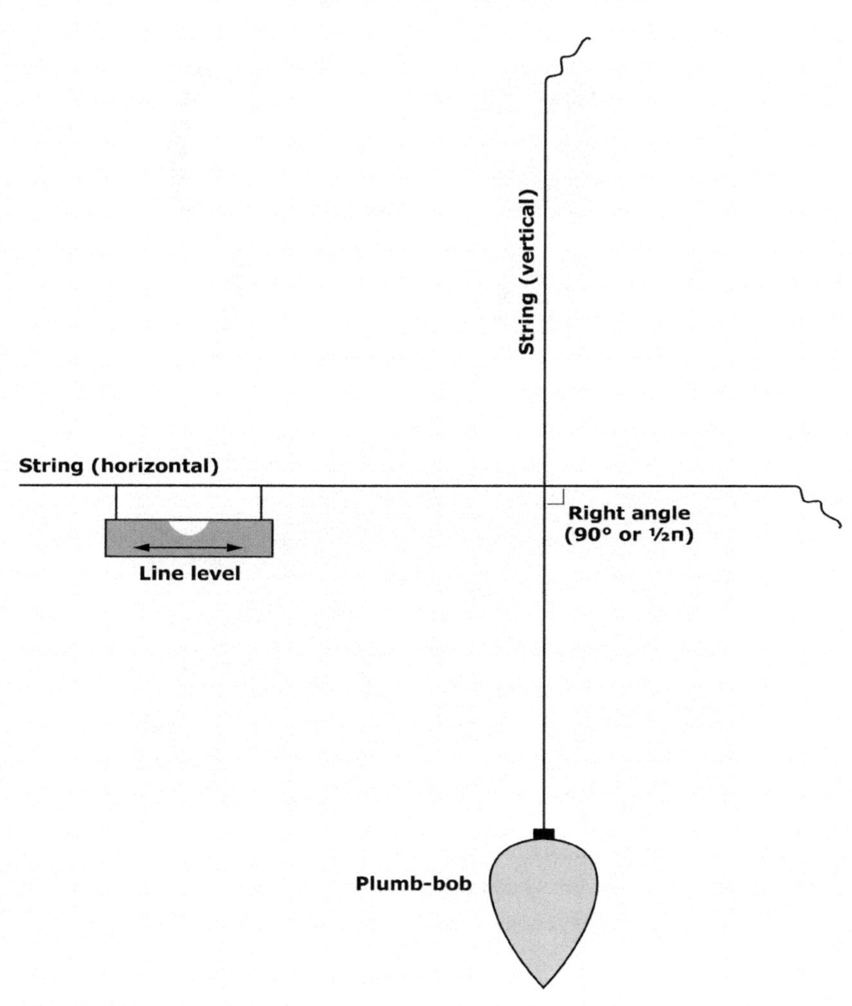

Figure 53 The axes for a Cartesian coordinate system in a vertical plane can easily be created by two strings: one kept horizontal with the aid of a line level, the second kept vertical with the aid of a plumb bob. Coordinates within this grid can be measured directly with a tape measure; see Figure 52, bottom, and Figure 64

the instrument; too low or too high will quickly tire the operator. This is not a concern if an automatic instrument is used. It is also important to leave some length of each of the legs, allowing them to move up and down, which will be necessary while setting the instrument up. Next the instrument is taken out of its box and secured on top of the tripod. There is no need yet to switch it on, unless the instrument has electronic spirit levels or is self-leveling.

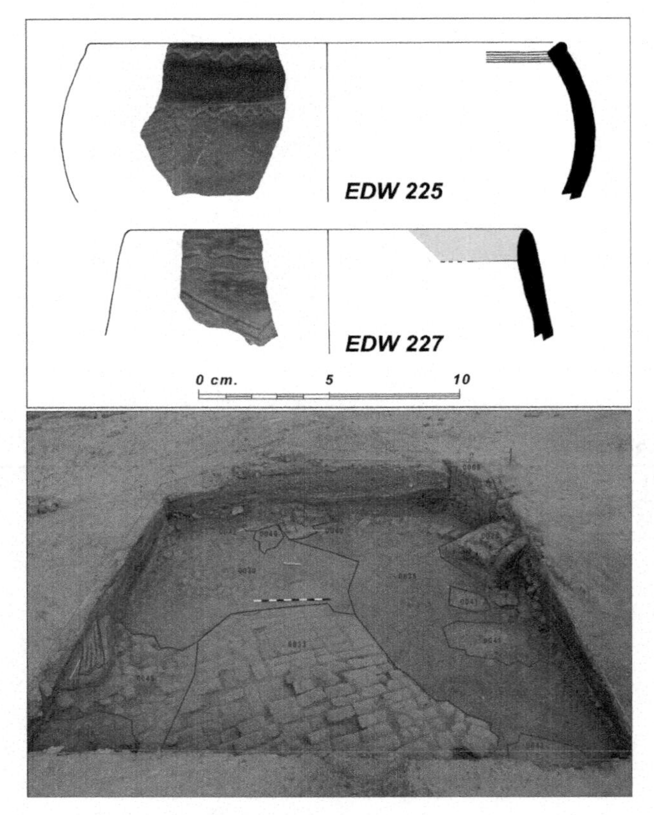

Figure 54 Standard technical drawing of two archaeological potsherds in which a photograph of the sherd has been inserted (top). Photograph of an excavation unit in which the layers have been drawn and indicated (bottom)

Table 14 Horizontal length of selected distances measured at selected slopes, see Figure 12, bottom. These figures are provided here to complement the text and illustrations

	5 m	10 m	15 m	20 m	25 m	30 m
5°	4.981	9.962	14.943	19.924	24.905	29.886
10°	4.924	9.848	14.772	19.696	24.620	29.544
15°	4.830	9.659	14.489	19.319	24.148	28.978
20°	4.698	9.397	14.095	18.794	23.492	28.191
25°	4.532	9.063	13.595	18.126	22.658	27.189
30°	4.330	8.660	12.990	17.321	21.651	25.981

Table 15 Length along the surface of selected slopes and horizontal distances; see Figure 12, bottom. These figures are provided here to complement the text and illustrations

	5 m	10 m	15 m	20 m	25 m	30 m
5°	5.019	10.038	15.057	20.076	25.095	30.115
10°	5.077	10.154	15.231	20.309	25.386	30.463
15°	5.176	10.353	15.529	20.706	25.882	31.058
20°	5.321	10.642	15.963	21.284	26.604	31.925
25°	5.321	10.642	16.064	21.284	26.604	31.925
30°	5.774	11.547	17.321	23.094	28.868	34.641

All total stations will have a device to find the point directly under the instrument. Often this is a prism (Figure 55), other instruments have a laser pointer or a hook for a plumb bob. To get the instrument in a level position over the desired point, first use the three leveling screws at the bottom of the instrument to move it to look directly at the correct position (Figure 55, bottom). Now that the instrument is directly over the occupied point, the final task is to level it. This is aided by the linear spirit level attached to the tribrach and moving with it. Usually there is a second, bull's-eye spirit level at the base of the instrument, which can mostly be ignored during this process.

Being on its tripod while aimed at the correct point, the instrument can be imagined to be at the surface of a large imaginary sphere of which the occupied point is the center. If the instrument can now be moved freely along the surface of this sphere, it should be possible to move it into the single position where it is both perfectly level while still directly above the correct point. To find this place, the three legs of the tripod should be made either longer or shorter (Figure 56, top). Therefore, turn the instrument so that the linear spirit level is in the plane of one of the legs, place one foot on that leg to prevent it moving too much, unlock the clasp holding the leg, and carefully lengthen or shorten it until the spirit level indicates that the instrument is horizontal in the chosen plane (Figure 57, top). Lock the leg in its new position and repeat this procedure for the other two legs, each time rotating the instrument so that the spirit level is in the plane of the leg that is worked on. Finally, repeat leveling the first leg. Now check if the instrument is nearly horizontal – which can be done quickly by checking the bull's-eye spirit level – and is still looking at the correct point. If the instrument moved too far from the correct point, use the leveling screws to move it back into the correct position and repeat the leveling procedure. If the instrument is not quite level, but

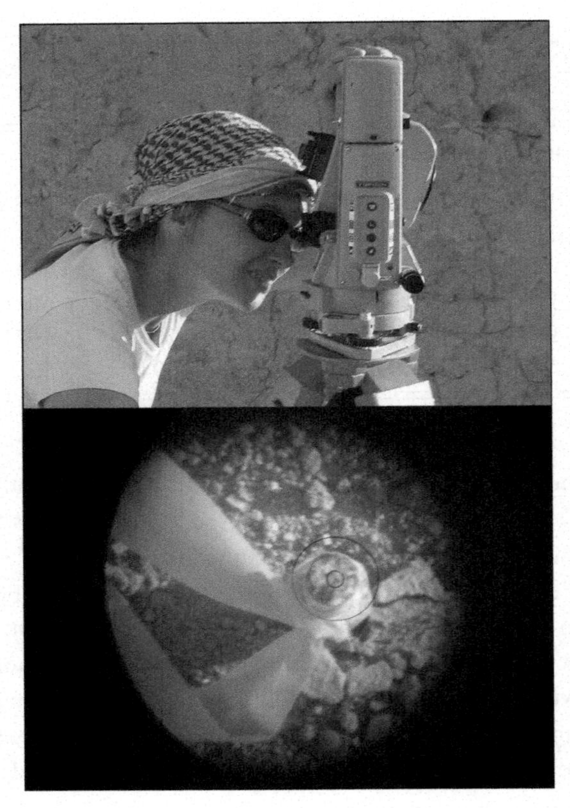

Figure 55 An archaeologist aims her total station, securely on top of its tripod, with the aid of the leveling screws (top), toward a grid point with known coordinates, marked by a length of rebar with flagging tape attached (bottom)

still looking at the correct point, keep lengthening and shortening the legs of the tripod until it is. This procedure might take some practice to master, but will soon become swift and effective.

Once the instrument is in the best position feasible, it should be turned so that the linear spirit level is in the plane effected by two of the three leveling screws. Now turn these two screws in opposite directions to move the spirit level into a perfectly horizontal position (Figure 57, bottom). Always move two screws at the same time and always move them in opposite directions (Figure 56, middle). Keep in mind that any movement of any of the leveling screws will move the instrument away from the correct point. Therefore, the better the instrument is leveled using the legs, the least it will have to be adjusted later. Once the instrument is level in the plane of two leveling screws, move the spirit level into the plane of the next two screws (one of which will already have been

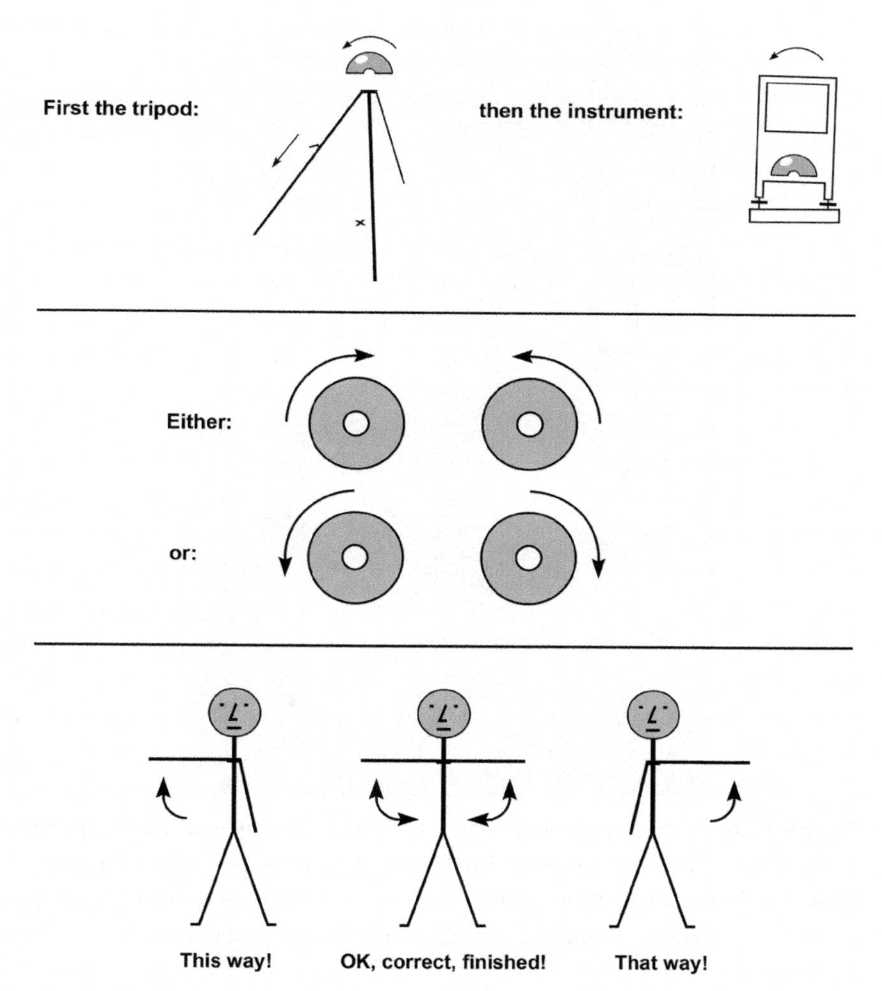

Figure 56 Basic instructions to level a total station in a desired position (top and middle). Basic hand signals to communicate with the person holding a stadia rod or prism target (bottom); see Figure 57

worked on). Level the instrument in this plane and repeat this procedure until the instrument is perfectly level.

Finally, check if the instrument, now in a perfectly horizontal position, is still looking at the correct point. If it shifted away a little, which is almost unavoidable, gently loosen the screw attaching the instrument to the tripod and gently slide the instrument into a position that it is looking at the correct point. Always repeat the procedure leveling the instrument with the leveling screws after this as the instrument will now no longer be perfectly horizontal. Repeat this

Figure 57 An archaeologist levels her total station directly above a grid point with known coordinates, marked with flagging tape, first by adjusting the length of each of the legs of the tripod (top), and subsequently by adjusting the leveling screws (bottom); see Figure 56 (top and middle)

procedure until satisfied. This too will take some practice to master, but will quickly become engrained.

The instrument is now ready for use and can be switched on. Most instruments are equipped with a prism suspended in the optical path, to help secure an accurate horizontal position of the telescope. If the instrument moves away from its horizontal position beyond the range of this prism, it will no longer return readings and will have to be adjusted. This usually happens by leaning on the tripod – especially when the instrument is set up too high or too low – accidentally kicking one the legs of the tripod, or by one of the legs sinking deeper into the ground. Automatic total stations are able to follow and find the user, who controls the instrument from the points to be located, rather than from the point occupied by the instrument and can be used after the instrument is set up properly.

Before taking measurements, make sure the telescope is in its upright position and not upside down. When used in an upside-down position, all positions will be 180° (π = 3.14159 radians) off in the horizontal plane, while the elevational differences will have the wrong sign (– will be + and vice versa). This is easily fixed in the final data set, but obviously better avoided. The final steps in setting up the instrument comprise focusing the optics and defining the horizontal 0° angle. The telescope of a total station has crosshairs in its optical path, similar to a level instrument (Section 2). It is better to work with the instrument without (sun)glasses. Fist the eyepiece of the telescope should be adjusted until the crosshairs pop into crisp focus. A different control on the telescope will focus the image of the landscape (Figure 16, bottom). The telescope can be moved along both its vertical and horizontal axes or locked in either or both directions.

To aim at the point to be located, the telescope should be free in all directions and moved until the desired point can be seen through the telescope. As the field of vision is small, most telescopes will have a simple sight on top to assist the observer in finding the target. Once the target can be seen through the telescope, the telescope should be locked in place, both horizontally and vertically, after which it can be moved precisely with two controls that slowly adjust the telescope along the horizontal and vertical planes. These four controls – two locks and two movement controls – are in different places on different instruments, but usually the horizontal lock and movement control are combined, as are the vertical lock and movement control. When first using a specific instrument, one should take some time to familiarize oneself and practice with the various controls.

When actually setting up the instrument, the first point to aim at is a backsight indicating the direction horizontal 0° angle (grid north) from the occupied point. There will be a way, either mechanically or electronically, to adjust the orientation of the horizontal protractor so that the horizontal measurements are taken as required. The orientation of the vertical protractor should have been established earlier, or can be so now, and usually cannot be changed. Once the instrument is fully set up, it can be used to locate as many unknown points (foresights) as necessary. Two different practical methods to set up a total station will be discussed here, after the description of the way to take individual measurements.

To take a foresight the telescope is aimed at a prism target on a pole directly above the point to be located (Figures 58 and 59). This should again be done by freeing the telescope, aiming it at the new point, locking it in its new position and adjusting it with the movement controls until the crosshairs are in the center

Figure 58 A prism target is used to reflect the electromagnetic beam of a total station, allowing it to calculate the distance between the telescope and the target (top); see Figure 59. Note that the lens of the camera is at the center of the prism and that this particular prism has a correction factor of –30 mm. The prism pole is kept vertical with the aid of an attached bull's-eye spirit level (bottom)

of the target. A total station measures angles based on the position of the telescope, which should therefore be aimed as accurately as possible, and distances by emitting an electromagnetic beam, which is reflected by the prism. The time difference or phase shift between the emitted and reflected signals allows the instrument to calculate the distance between the telescope and the prism. Using these measurements the instrument can calculate projected distances as well as coordinates in three dimensions (Figures 6 and 12). Some prisms require a correction factor to be entered into the instrument for distances to be calculated correctly (Figure 58, top). To minimize errors the prism target has to be held steady and directly above the point to be located. The pole should therefore be as short as possible and kept vertical with the aid of the attached bull's-eye spirit level (Figure 58, bottom). Some total stations can be used without a prism target and measure angles and distances to any point on a sufficiently reflective surface.

Figure 59 Archaeologists locate ancient remains using a total station and prism target; see Figures 46, top, and 65. Note the marked grid point below the total station, see Figure 55, and the vertical position of the prism pole, see Figure 58

It has to be kept in mind that elevations are calculated from small angular differences, and even smaller differences in the cosines of these angles, over relatively long distances and may therefore be less accurate than elevations measured directly with a level instrument (Section 2). Obviously the height of the instrument and the height of the target have to be entered into the instrument to be taken into account during its calculations (Figure 59). All instruments are capable of storing the raw data and show the operator either polar coordinates (angles and distances) or the three-dimensional coordinates (x, y, z) of the points measured, and often also visualize the measured points. This allows for large data sets to be collected swiftly and accurately. Automatic total stations are able to locate the prism target and have the controls attached to the prism pole, allowing the survey work to be done by a single person.

Despite the storage and visualization facilities of modern instruments, it remains vital to keep field notes in order to control the work flow, both during and after the fieldwork, and maintain an overview of the various data streams (Figure 60). These notes should preferably include a sketch map of the area surveyed, the name of the file in which the associated data is stored, the coordinates (or identifying number) of the occupied point, the coordinates (or identifying number) of the backsight, the height of the instrument above the

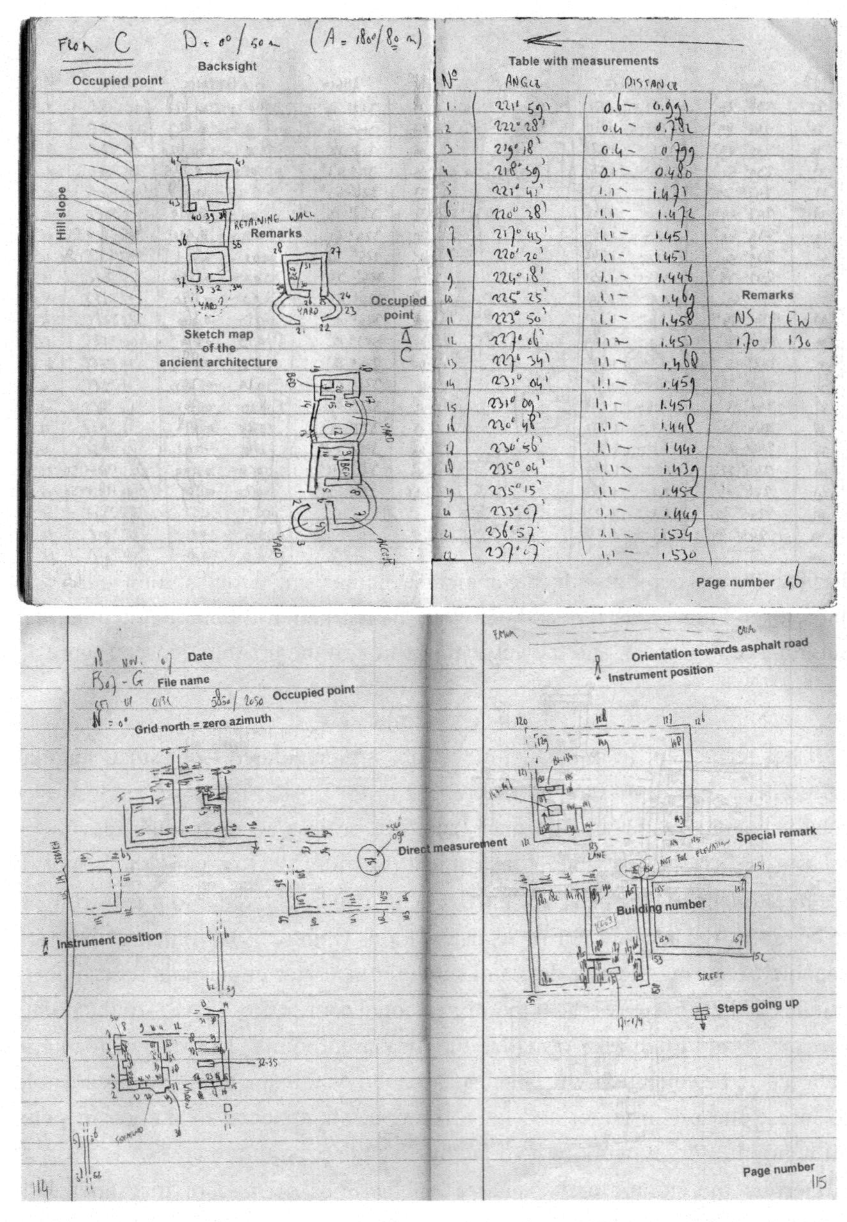

Figure 60 Two annotated pages from an archaeological surveyor's notebook; see Figures 2 and 65

ground, the height of the prism target above the ground, and any additional remarks or interpretations that will not readily be evident from the collected data. A second vital point during each survey session is to locate a few points with known coordinates as this will allow the data to be checked and corrected, if necessary. If the same points are recorded at both the beginning and end of the session, this allows possible drifts in the data to be identified and possibly addressed.

Three final issues are worthy of attention here. The effect of the instrument being used with the telescope upside down and the fix for the erroneous data has already been discussed. Simply add 180° (or 3.14159 radians) to the horizontal angles and change the sign of the differences of the vertical angles with the horizontal plane (which is not necessarily 0°) to make the points fall in the right place. Second, if the instrument suddenly stops taking measurements, without the battery being drained, it is most likely no longer level. As this will slightly change both the occupied point and the horizontal 0° angle, this means that the instrument has to be set up as if newly placed in its current position. This should be relatively swift as the instrument is unlikely to have moved very far. Third, after completing the survey the instrument has to be switched off, the leveling screws returned to a central position, and the instrument placed back in its box with the horizontal and vertical movement left unlocked. If allowed to move freely, within the confines of the padding in the box, damage to the mechanical parts of the instrument is less likely to occur. Attention must be paid to the way in which the instrument fits snugly in its box. Marking the positions of the moving parts relative to each other will facilitate taking down and storing the instrument.

Apart from locating points with unknown coordinates, archaeological survey sometimes entails placing markers in the terrain at predetermined coordinates. With the instrument set up in a known location, it is easy to guide the person holding the prism pole to the imaginary line between the desired location and the instrument. For this the telescope of the instrument is placed at the correct angle and the pole followed visually until it meets the crosshairs in the telescope. As the distance between the person at the instrument and the person with the pole increases, communication between them may become difficult. When they do not have access to a cell phone or a walkie-talkie, it is important to agree on simple and clear messages, either verbal or by using hand signals (Figure 56, bottom). Remember that the person at the instrument is aided by the optics of the telescope, while the person with the pole is not. Sometimes it will prove necessary to enlarge the hands giving the signals with a notebook or a hat, or to step away from the instrument to allow a better view, unobstructed by the instrument on its tripod. Another obstacle to unambiguous communication is the use of abbreviations, the most common of which are explained in Table 16.

The correct distance is more difficult to estimate, and the use of a measuring tape to make a first estimate will prove helpful. It will prove useful for

Table 16 Selection of abbreviations often used during archaeological survey work

BM	bench mark (local datum)
BS	backsight
COS	cosine
D.ddd	decimal degrees
DGPS	differential GPS
DEM	digital elevation model
DMS	degrees–minutes–seconds
EPE	estimated position error
ETA	estimated time of arrival
FS	foresight
GIS	geographic information system
GPS	Global Positioning System
HD	horizontal distance
HI	height of the instrument (day height)
HM	height of the mirror (prism target)
LiDAR	light detection and ranging
mASL	meters above sea level
OCC PT	occupied point
RAD	radians
SBAS	Satellite-Based Augmentation System
SD	slope distance
SIN	sine
SQRT	square root
SS	side shot
SSM	standard survey monument
UAV	uncrewed aerial vehicle (drone)
UTM	Universal Transverse Mercator system
VD	vertical distance
WGS84	World Geodetic System 1984 (ellipsoid and datum)
ZEN	Zone–Eastings–Northings

any surveyor to know the length of her or his stride, allowing distances to be assessed fairly confidently by counting paces. The correct distance is obviously established with the aid of the total station. Once the first point is laid out, the following are easier to estimate by the person holding the pole as each new point should line up with those laid out earlier as well as the instrument.

There are many ways to design a measured survey, partly depending on the local situation and the work already done at the site, but a practical method that works in many circumstances is briefly described here. All missing details can be gleaned from the previous sections, if and when necessary. First find a convenient spot for the total station, roughly in the center of the site and with a clear view of most of it. Set up the total station so that it is in a stable position and perfectly level. Determine the location of the instrument with a (D)GPS device, set to UTM coordinates (Section 4), and note this down. Now walk away from the instrument while viewing the display of the (D)GPS device, keeping the eastings stable and the northings increasing. After pacing out 100–200 m allow the device to settle for a while and leave a physical marker almost due north of the point occupied by the total station (Table 11). Even when using a handheld, uncorrected GPS device, this creates an acceptable approximation of a small section of the meridian on which the total station has been placed. Back at the instrument the telescope is aimed at the marker and the horizontal protractor set to 0°. Now all measurements made will fall within the UTM grid and can readily be combined with data collected by a GPS device or extracted on existing maps.

Next, at least three more markers should be left in the terrain in a way that ensures that they will remain in place for the duration of the survey. Often it is sufficient to hammer lengths (30–50 cm) of rebar – which is not expensive and readily available in most hardware stores worldwide – into the ground (Figures 57, top and 58, right). These markers should not be in a straight line or on the circumference of a circle. As at least three markers should be visible from anywhere on the site, it may be necessary to leave more than the four mentioned here. After leaving a sufficient amount of markers the coordinates of the occupied point, established with the GPS device, are entered into the total station and the four or more markers located. The coordinates of these, which should be stored in the memory of the instrument as well as written down in the field notebook, will be used to locate and orient the instrument during the remainder of the survey work with a method called resection. Even though the location and orientation of the resulting map may be slightly off, depending on the EPE of the occupied point and the difference between grid north and true north, it will be internally consistent and can be amended when more accurate geospatial information becomes available (Figure 41). After setting up in this way, the actual survey work can commence and data on the visible ancient remains collected.

For the next session of survey work the total station can be set up in any convenient place from which at least three of the markers with known coordinates are visible, using a method called resection (Figures 61 and 62). This technique will be embedded in the software of the total stations and provides the most accurate way of locating and orienting the instrument. Once the

The Tienstra solution for resection:

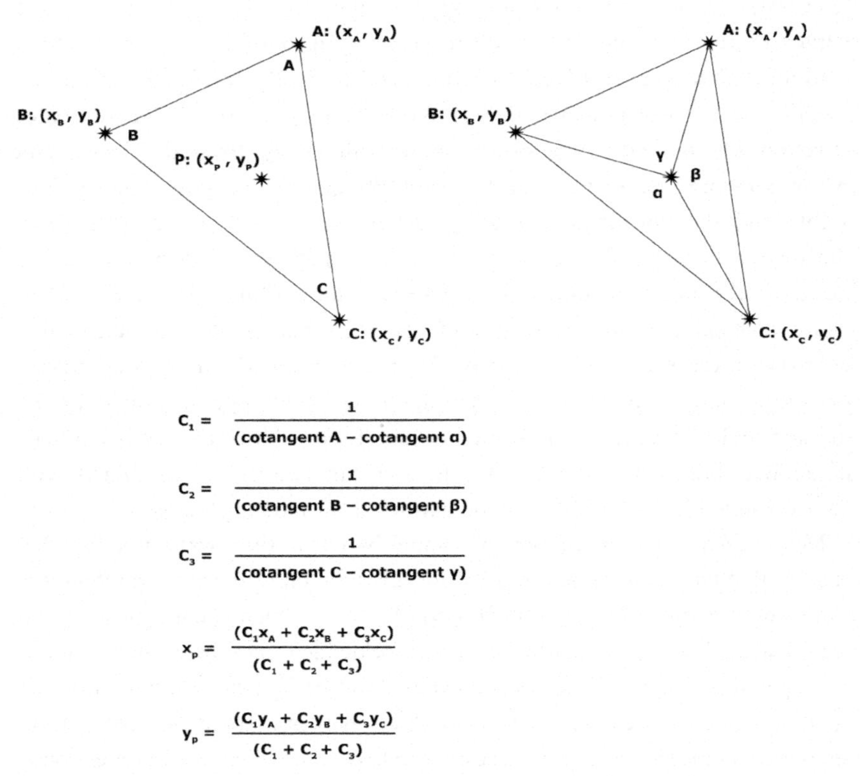

Figure 61 The Tienstra solution for finding the position of point P (the point occupied by the survey instrument) using only the known coordinates of points A, B, and C and the angles between the six lines connecting A, B, C, and P (resection). This or a very similar algorithm is embedded into the software in total stations and provides the most accurate way of locating the point occupied by the instrument within a previously defined grid; see Figure 62

program has been started the instrument will give instructions on how to proceed. This will involve locating two of the markers, after which a provisional position can be calculated, which can subsequently be refined by locating additional markers. It is advisable to locate at least three or four markers, after which the gain in accuracy diminishes quickly. The algorithm will not work for markers that are in a straight line or on the circumference of a circle. Although if can be finicky at first, setting up will become a matter of minutes once the layout of the site and the survey area becomes more familiar.

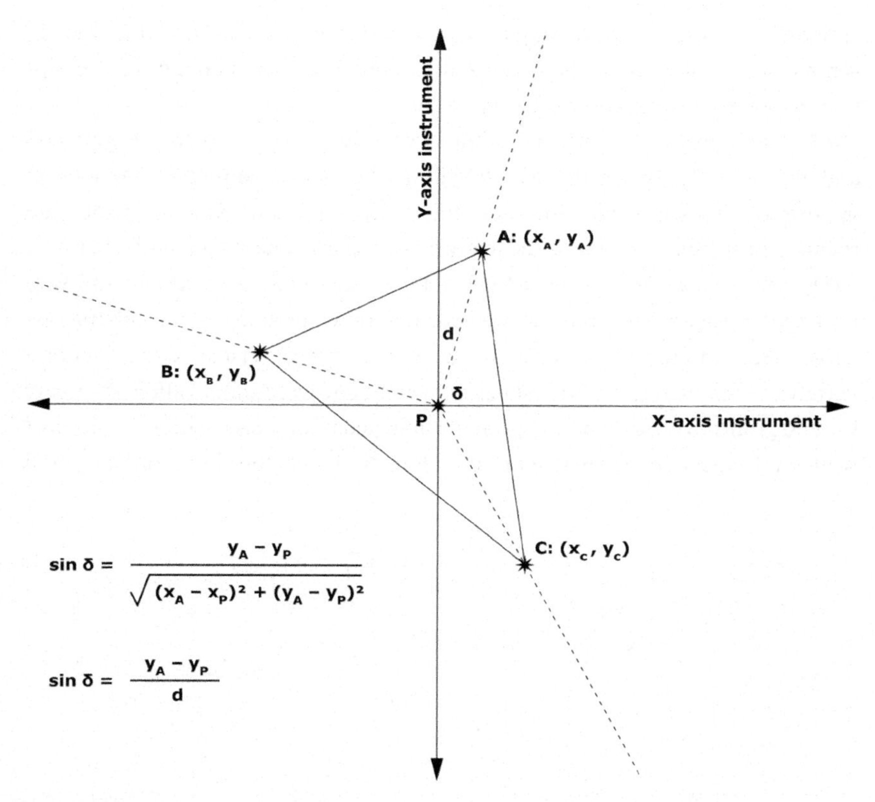

Figure 62 Once the coordinates of the occupied point P are known, see Figure 61, the azimuth toward grid north (= 0° in the horizontal plane) can be calculated using the coordinates of points A, B, C, and P. This or a very similar algorithm is embedded into the software of total stations and provides the most accurate way to orient the instrument within a previously defined grid

6 Epilogue: Data Reduction

The previous sections provided an overview of the basic methods and techniques to collect raw geospatial data on archaeological materials and remains. The next step will be the reduction of these data sets – mostly consisting of large series of numbers – into maps, plans, graphs, or any other readily comprehensible format. This process is discussed here only in general terms because of the rapidly developing selection of options, many of which entail proprietary software, and the changes that are frequently made to most of these. As it is near impossible to keep up with these advancements, also for financial reasons, most archaeological surveyors have developed personal preferences on how to approach the issue of visualizing the data that they collected. Therefore, mostly samples of finished projects are presented here, together with a brief description, to serve as illustration and inspiration.

Plans and profiles of excavation units are often drawn by hand (Figures 63 and 64), not only for reasons of accessibility but also because this provides an opportunity to study and interpret the ancient remains and their intricate relationships (Section 5). For the analysis of the geospatial data collected with a (D)GPS device (Section 4) or total station (Section 5), a variety of software packages are available. Some are dedicated to downloading and visualizing raw survey data, to create contour maps, or to analyze and visualize the data in a variety of other ways. At its most basic level the software should allow the many located points to be shown to scale and connected with lines to show the faces of walls and other structures (Figure 65). For the latter, the field notebook will

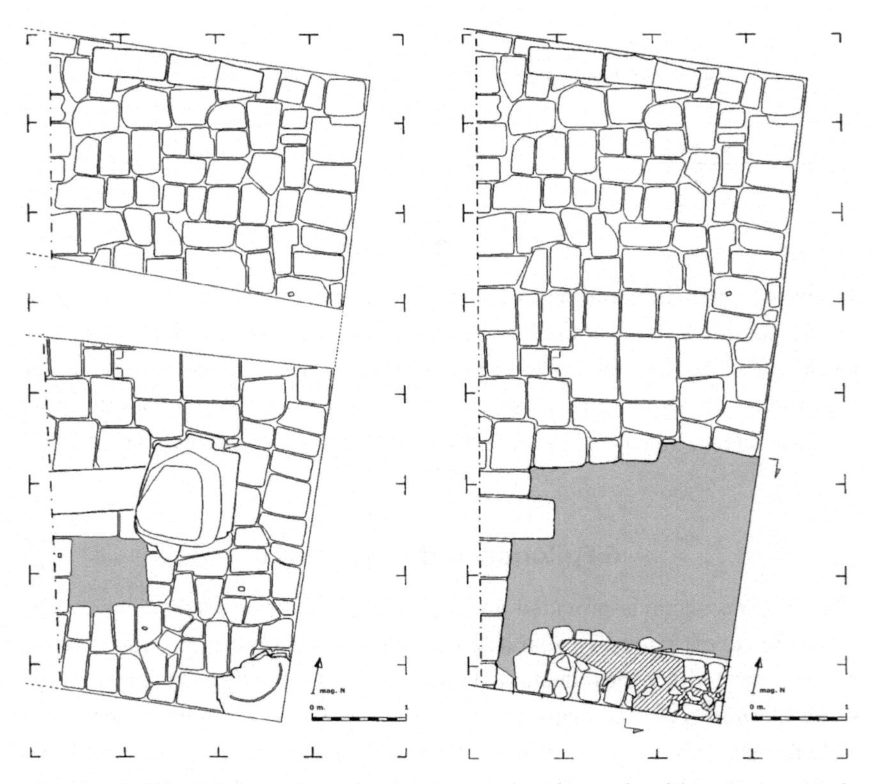

Figure 63 Hand-drawn plan of a 4 × 8 m excavation unit with a stone-paved floor, a stone bench (top), the remains of a column base (bottom-left), and an older structure below the pavement (bottom-right), showing two different stages of excavation; see Figure 51

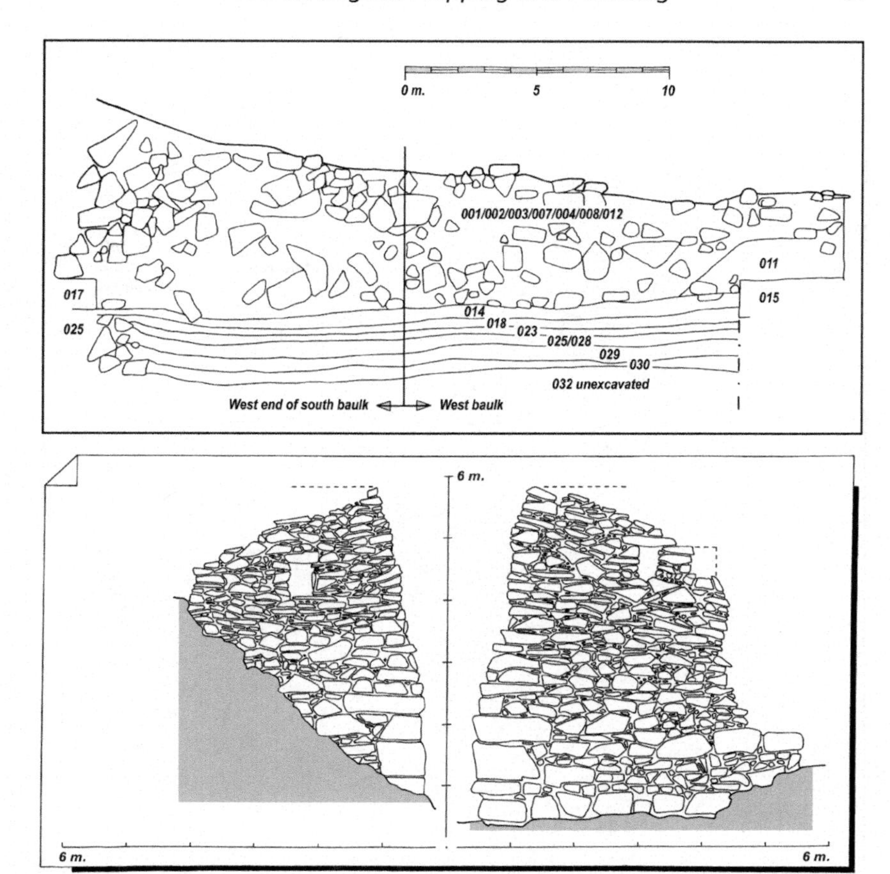

Figure 64 Hand-drawn profile of the western and southern profile (baulk) of an excavation unit (top), with the different stratigraphic layers indicated, and the hand-drawn elevation of two faces of the corner of an ancient stone building (bottom), built into a steep hillslope (see Figures 52 and 53)

prove indispensable (Figure 60). Advanced data loggers allow the process to be done in the field – thus essentially functioning as a field notebook – and the results to be downloaded later.

Many software packages exist that can visualize numerical properties connected to a two-dimensional geospatial data set (x, y). Additional properties include the elevation of each point (Section 2), but can also be the pressure of the air, the salinity of the ocean, the frequency of the occurrence of the northern lights, or the density of ancient pottery sherds across the landscape. Visualization of such data sets is often in the form of a contour map (Figures 20 and 66), a digital elevation model (Animation 4), or a heat-map (Figures 22, bottom and 67). Every

Figure 65 Section of the plan of a large archaeological site created with a total station and basic data reduction and visualization software; see Figures 59 and 60

property that can be reduced into a numerical value may be visualized this way. Connecting more or less complex databases to one or more maps is referred to as creating a geographic information system (GIS). Pioneers of this idea include Edmund Halley (Section 2), Charles Joseph Minard (1781–1870), who in November 1869 famously created a map of the successive losses of men in the French Army during its Russian campaign of 1812–13 (*Carte figurative des pertes successives en hommes de l'Armée Française dans la campagne de Russie 1812–1813*), among many other similar infographics, and John Snow (1813–58), who plotted the cholera epidemic of 1854 on a map of London, showing that the

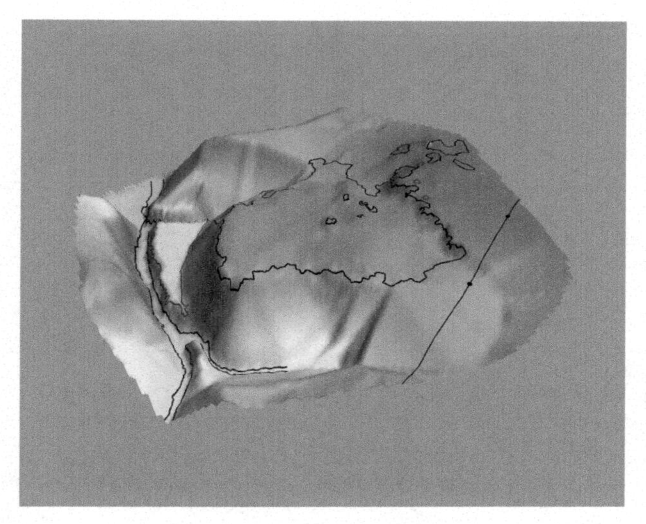

Animation 4 Rotating digital elevation model of an archaeological site and its environs. The animated version of the image is available at www.cambridge.org/barnard

contaminated water of the Broad Street pump had led to a localized outbreak of cholera. The ultimate intent of creating a GIS, apart from simply storing and collating the data, is for heuristic use, meaning to obtain insights which were not obvious before in ways similar to, for instance, John Snow's map or the coxcomb diagrams created by Florence Nightingale (1820–1910).

Often newly collected, archaeological data need to be combined with existing maps, aerial photographs, or satellite imagery, which does not always have the necessary geospatial information digitally embedded. Most visualization software packages enable imagery to be georeferenced to make it fall in the right place. To do so, four to twelve points on the image need to be identified and their coordinates established in the field as accurately as possible. On photographs this can be done by locating points visible in the image with the aid of a total station or a (D)GPS device. On maps it can usually be achieved by identifying the crossings of the meridians and parallels. Once the coordinates of a sufficient number of points – as evenly distributed across the image as possible – are known, the program can use these to orient the image in the virtual three-dimensional space in which the archaeological data is or will be collected. Depending on the way in which the image was produced and the projection chosen for the final result, images may appear distorted after this process. Subsequently, several different images can be placed on top of each other and the archaeological data projected on any of these (Figure 68).

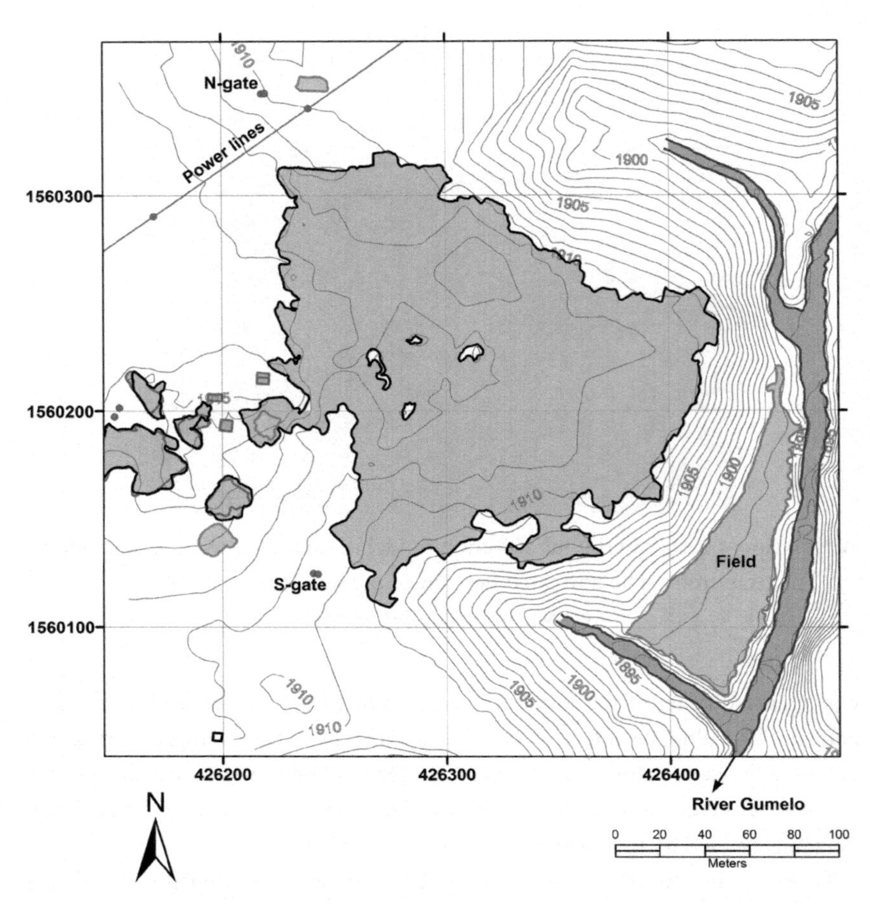

Figure 66 Computer-generated contour lines, based on thousands of DGPS data points, see Figures 39 and 40, overlaid on the schematic map of a large archaeological site, see Animation 1

Apart from satellite imagery, much aerial photography is now taken from an uncrewed aerial vehicle (UAV), commonly referred to as a drone. Most drones are equipped with a GPS receiver, but the imagery often comes without geospatial data embedded and needs to be georeferenced after being downloaded from the device (Figure 68). As the time and place of flying a drone is under the control of the archaeologists, it is possible and advisable to place markers in the area to be photographed and locate these before or after the pictures have been taken (Figures 69 and 70). A dozen markers left in strategic positions will result in very accurately located imagery. Some drones come with an embedded GPS system and software that enable them to fly in programmed patterns. The resulting imagery can

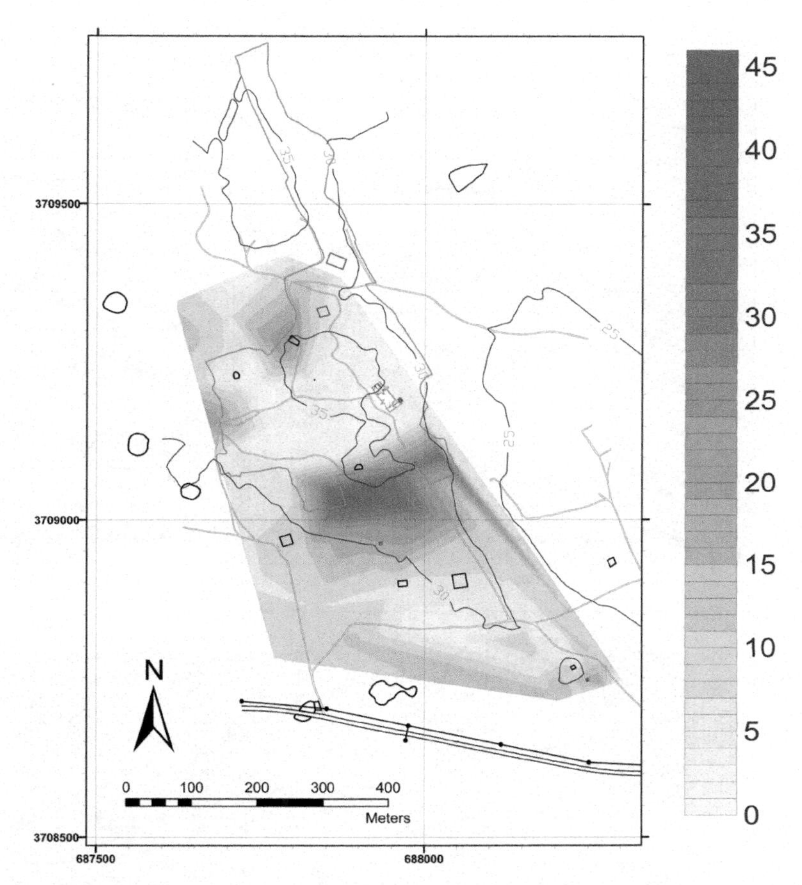

Figure 67 Heat-map showing the density of pottery sherds found on the surface of a large archaeological site; see Figure 22, bottom

be projected onto a horizontal plane and allow for the inference of contour lines. More expensive options include mounting an instrument that can create a three-dimensional scan of an archaeological site onto a drone or aircraft, a technique referred to as LiDAR (light detection and ranging). This allows the investigation of large areas and is not significantly hampered by a canopy of leaves.

A final technique with increasing importance in archaeological surveying is structure-from-motion, often referred to as photogrammetry, which entails the creation of three-dimensional models out of a large number of photographs (Models 1 and 2, SHIFT-click allows zooming in and out, CONTROL-click allows to pan). To create such models a large number of photographs need to be taken (Figure 71, top), covering all visible surfaces and with sufficient overlap for the program to create a high-resolution model. To facilitate this and to

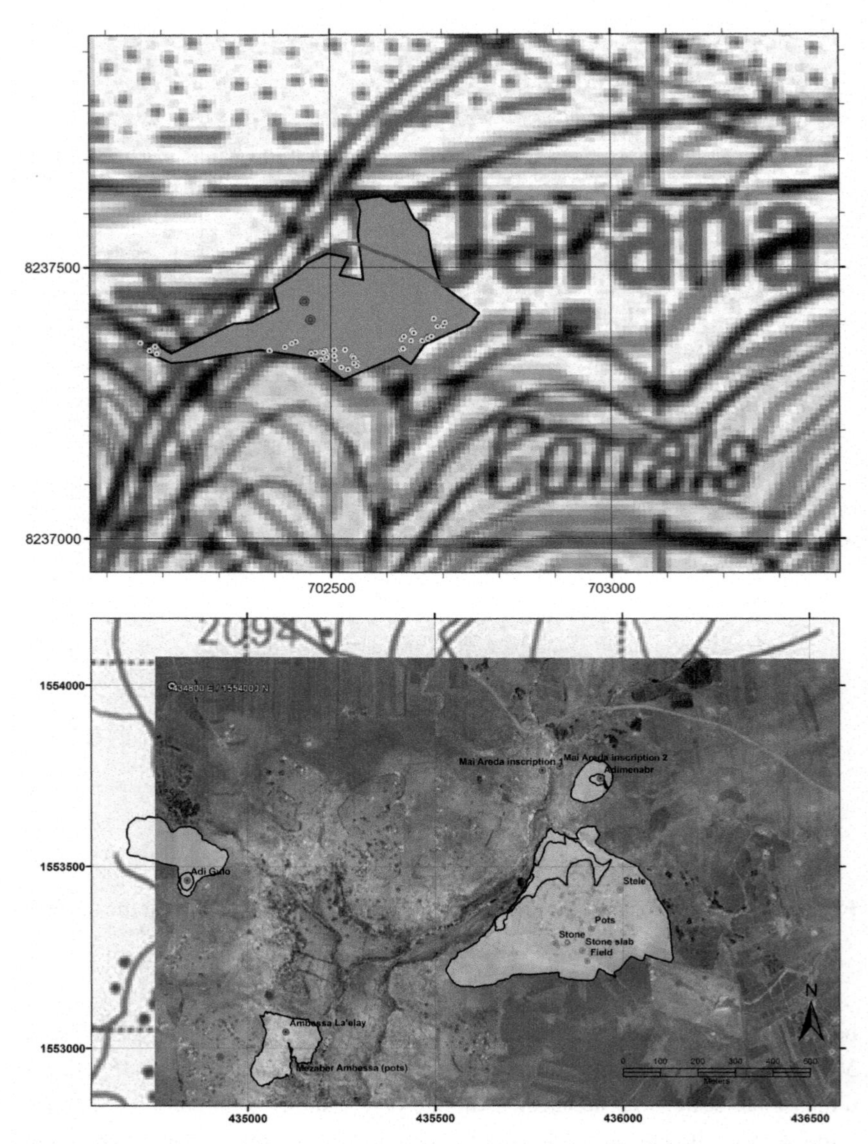

Figure 68 Archaeological geospatial data overlaid on an existing topographic map of the area (top), and archaeological geospatial data overlaid on both a topographic map as well as a georeferenced satellite photograph of the area (bottom)

georeference the final result it is necessary to attach a number of targets in strategic places (Figure 71, bottom). Depending on the size and complexity of the structure or object to be modelled, it may be necessary to capture 50–500 photographs. For better results efforts should be made to have limited contrast in the original photographs, and either cast uniform shade or collect the imagery in the early morning and late afternoon. The remote background should also be avoided or removed digitally before the imagery is processed, as this may confuse the software.

Figure 69 In preparation of a drone survey of an archaeological site, a number of targets are placed on the ground (top). Section of the resulting orthorectified image in which a number of such targets can be seen (bottom). *Most countries have regulations for the use of drones. Always check with the local authorities before flying one.*

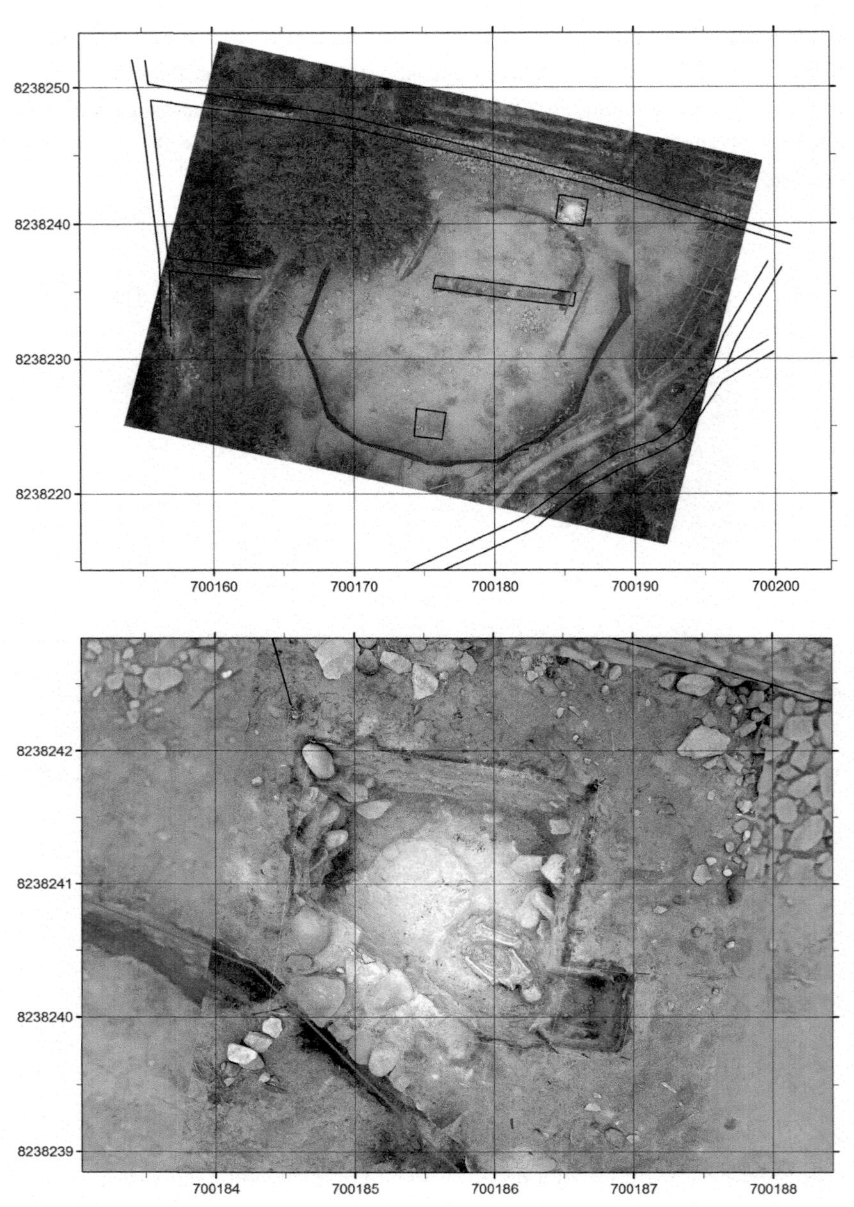

Figure 70 Two georeferenced photographs of the same archaeological excavation site, taken from an uncrewed aerial vehicle (drone), overlaid on the initial topographic map of the site. *Most countries have regulations for the use of drones. Always check with the local authorities before flying one.*

Model 1 Digital three-dimensional model of the excavated remains of a Roman bath house. The 3D version of the image is available at www.cambridge.org/barnard (click to explore, SHIFT-click to zoom in or out, ALT-click to pan)

Model 2 Digital three-dimensional model of the excavated remains of a Roman granary. The 3D version of the image is available at www.cambridge.org/barnard (click to explore, SHIFT-click to zoom in or out, ALT-click to pan)

Figure 71 An archaeologist takes photographs to combine into a three-dimensional digital model of a recently excavated ancient building (top); see Models 1 and 2. Sample targets as used to facilitate the stitching and georeferencing of photographs (bottom)

Once uploaded into the program, this will first select a large number of sections that seem to appear in more than one image. This process is assisted by the targets. These points are subsequently connected by lines, resulting in a wire-frame surface consisting of triangles. This step explains the term structure-from-motion, in which the structure of an object is inferred from the motion of the camera around it. Finally, the surface is clad with the relevant sections of the photographs to recreate the appearance of the original. Most software is able to fill small areas where insufficient data is available. In places where no data could be collected, such as below the surface of an excavation unit, the program will either display a mirror image of the actual surface or nothing at all. The whole process uses a significant amount of computer processing power and thus requires a large working memory, a good graphics card, and a significant amount of time (large models may take 4–24 hours to complete).

The models are obviously best viewed using the software environment in which they were created. Some allow the angle of the incoming light to be moved around, thus enabling the study of the modelled object in raking light coming from various directions. When properly georeferenced the model can be scaled and placed in a virtual three-dimensional grid in which measurements can be taken. Doing so during various stages of work in an excavation unit allows the depositional history and stratigraphy of the unit to be more or less accurately reconstructed and studied. This comes close to the wish, mentioned at the very beginning, to be able to reconstruct the archaeological deposits and repeat their excavation. Most programs also allow the model to be projected onto a given plane, most often either vertical or horizontal (orthographic), to complement a profile drawing or a plan (Figure 72). To share models with those that do not have the software in which it was created, including in an online environment, most programs allow models to be exported in a .pdf-format, with the necessary software to manipulate the model embedded (Models 1 and 2).

Much of the collection and reduction of archaeological data is increasingly moved onto digital and online platforms, partly facilitated by the ubiquity of mobile telephones, a trend which ensures less chance of errors and greater ease of accessing, searching, storing, transporting, copying, and distributing data. At the same time it facilitates the analysis of large data sets and enables meta-analysis, which comprises comparing data sets compiled by different research groups at various times and places. Only partly resolved issues with this change, however, are questions of ownership and curation of the data. It is hard to protect the copyrights of freely accessible, digital online data and secure that (academic) credit is awarded where this is due. Sensitive archaeological data on sites and objects can be used by looters and traders in antiquities to advance their criminal activities. Furthermore, the large variety of archaeological data, spanning large time frames and significant diversities in cultures, seems to render a universal standard layout of archaeological data systems implausible. These issues are less pertinent for geospatial data and using a GIS platform, linking archaeological data to its geographic location, may provide a basic model for many archaeological data storage projects.

Another major, mostly unresolved issue with digital archaeological data, both off- and online, is the rapid development of both the software and the hardware needed to access and manipulate the data. On the one hand, these developments constantly increase the speed, accuracy, and convenience of the used systems,

Figure 72 Orthographic aerial views of two excavation units created by projecting a three-dimensional model onto a horizontal plane; see Models 1 and 2

but on the other hand, they often render older data sets nonfunctional. This necessitates a constant curation of existing data sets, which at times may need to be converted or migrated into dissimilar systems in order for them to remain functional, which can be labor intensive and expensive.

In addition to the issues briefly described here and the fact that scholarly publications are still modeled on traditional printed media, which do not readily allow the inclusion of interactive maps or digital models, a final issue is caused by the dynamic nature of digital information. Records, files, and documents can be amended or removed without this being immediately obvious, rendering references made to older versions incorrect. Appropriate consideration should therefore be given to version control – the acknowledgment and accessibility of previous versions of digital records, files and documents – and to link rot, the phenomenon that connections between records, files, and documents will get lost if they are moved or renamed.

All the problems discussed in this section can be addressed and resolved, but it has to be kept in mind that much digital data collected and stored ten to fifteen years ago using now-obsolete hard- or software is no longer serviceable, while the maps created on paper by George Washington (1732–99), who worked for a short time (1749–50) as surveyor for Culpepper County before becoming a general and the first president of the United States – and who continued to do survey work until a few weeks before his death – have remained readily accessible and comprehensible. The same is true for the elements of archaeological mapping and planning presented here.

Further Reading

Banning, Edward B. (2002), *Archaeological Survey*, New York (Kluwer Academic Press).

Collins, James M. and Brian Leigh Molyneaux (2003), *Archaeological Survey: Archaeologist's Toolkit Volume 2*, Walnut Creek (AltaMira Press).

Kavanagh, Barry F. and Tom B. Mastin (2014), *Surveying: Principles and Applications: 9th edition*, Boston (Pearson).

McCormac, Jack C., Wayne A. Sarasua and William J. Davis (2013), *Surveying: 6th edition*, Hoboken (John Wiley).

Nathanson, Jerry A., Michael T. Lanzafama and Philip Kissam (2018), *Surveying Fundamentals and Practices: 7th edition*, New York (Pearson Education).

White, Gregory G. and Thomas F. King (2016), *The Archaeological Survey Manual*, Milton Park (Routledge).

Cambridge Elements

Current Archaeological Tools and Techniques

Hans Barnard

University of California, Los Angeles

Hans Barnard is an Associate Adjunct Professor in the Department of Near Eastern Languages and Cultures at the University of California, Los Angeles, as well as an Associate Researcher at the Cotsen Institute of Archaeology.

Willeke Wendrich

University of California, Los Angeles

Willeke Wendrich is the Joan Silsbee Chair of African Cultural Archaeology and the Director of the Cotsen Institute of Archaeology at UCLA. In addition she is Professor of Egyptian Archaeology and Digital Humanities in the Department of Near Eastern Languages and Cultures at the University of California, Los Angeles as well as the Editor-in-Chief of the online UCLA Encyclopedia of Egyptology.

About the Series

Cambridge University Press and the Cotsen Institute of Archaeology at UCLA collaborate on this series of Elements, which aims to facilitate deployment of specific techniques by archaeologists in the field and in the laboratory. It provides readers with a basic understanding of selected techniques, followed by clear instructions how to implement them, or how to collect samples to be analyzed by a third party, and how to approach interpretation of the results.

Cambridge Elements ☰

Current Archaeological Tools and Techniques

Elements in the Series

Archaeological Mapping and Planning
Hans Barnard

A full series listing is available at: www.cambridge.org/EATT

Printed by Printforce, United Kingdom